More Than Free Gold

To write a book on Canadian mining requires industry ...

Arnold Hoffman, *Free Gold*

Mineral Exploration in Canada Since World War II

MICHAEL BARNES

GENERAL STORE PUBLISHING HOUSE
499 O'Brien Road, Box 415
Renfrew, Ontario, Canada K7V 4A6
Telephone (613) 432-7691 or 1-800-465-6072
www.gsph.com

ISBN 978-1-897113-90-5

Printed by Custom Printers of Renfrew Ltd., Renfrew, Ontario
Printed and bound in Canada

Design and layout: Mag Carson

Library and Archives Canada Cataloguing in Publication data:

Barnes, Michael, 1934-
More than free gold / Michael Barnes.
ISBN 978-1-897113-90-5
1. Mineral industries--Canada--History--20th century.
I. Title.
HD9506.C32B37 2008 338.20971'09045 C2008-906347-0

Cover photos:

Front cover, background photo: Redpath workers at the Kidd Creek Mine (*G. Oxby*)
Front cover inset photos, left to right: Diavick Mine, N.W.T. (*Jiri Herman, Diavik D.M.*); Plane picking up sample at Hackett River (*Sabina Silver*); Ikonos 2 satellite (*NASA/Wikimedia Commons*); Ship in northern waters (*Iron Ore Company of Canada*); Students at Northwest Community College, Smithers, B.C. (*NWC College*); Geophysicist works in the west Raglan area, Labrador (*Knight Resources*)

Back cover, top to bottom:
Fisheries biologist Genevieve Morinville at EKATI Diamond Mine (*Rider Petch*); Eira Thomas with her brother at the site of a kimberlite discovery in the mid 1990s (*Stornoway Diamonds*); Visitors to the Diavik mine.

This book is for Robert Cudney and John Harvey
who were instrumental in seeing that this story
of more than six decades of exploration in Canada
was preserved for future generations;
and Dr. Jim Franklin who provided invaluable assistance
and generously wrote the commodity chapter introductions.

WORTH NOTING

All proceeds from the sale of *More Than Free Gold* by the Prospectors and Developers Association of Canada will be contributed to Mining Matters, the education arm of the PDAC, a registered charity which promotes among students an awareness of Canada's geology, mineral resources and mining industry.

Contents

Author's Note

Those who decry mining are ignorant of history. If they knew anything about metals, they would know that all business, all industry and all human progress depend on mines.

William Sulzer, Governor of New York, 1938

Drawing by Irwin Hoffman

One of the guiding lights for anyone interested in the mining industry of Canada and what came before is Arnold Hoffman's book *Free Gold: The Story of Canadian Mining*. Published in 1947, this work might at first glance be considered an odd candidate to have gathered the following it has acquired in the more than sixty years since it first appeared. Written by an American geologist who was raised in Boston and never saw the bush until he came to Canada after World War I, it nonetheless has captured a large following because of the writer's obvious deep feeling for Canada and those involved in its mining industry.

Arnold Hoffman

Irwin Hoffman

Cover of the 1947 edition of *Free Gold: The Story of Canadian Mining*

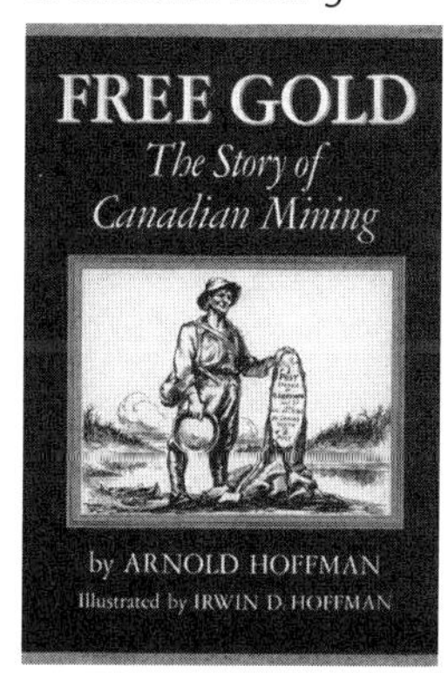

Along with his brother Bob, Arnold Hoffman prospected mostly in Quebec and Ontario but he visited all the major mining camps in the country and met many of the figures who became prominent in the industry. His gift, though, was that he could tell the story not only of the spectacular achievers but also of the average prospector, geologist, promoter and developer from their own points of view. His writing was engaging and never dull. Another brother, Irwin, who later became a prominent painter in the United States, enhanced the book with sketches of mining men.

The present author found *Free Gold* to be a major signpost for several of his books relating to mining in Ontario. During work on *The Scholarly*

Dr. Jim Franklin
San Gold Corp.

Prospector, the story of Don McKinnon of Hemlo fame, I met John Harvey, a geologist, mine finder, and more recently, a consultant to the industry. John was convinced that there should be a book which would take up the story where Arnold Hoffman left off — hence the present title, *More Than Free Gold*. This work would be the story of exploration in Canada over the past sixty years. He went further and laid out the road map for the content and direction of this modest sequel. The project would have to be major in scope but, of necessity, relatively brief in finished form. Dr. Jim Franklin, a prominent geologist and former chief scientist of the Natural Resources Canada team, joined John Harvey as a mentor for this project. His major contribution was to provide advice, create graphs and formulate introductions to the commodities.

Such a work needed an angel — someone who saw the need for the mineral exploration story to be told and who had the wherewithal to see it brought to fruition. Such a man was Robert Cudney of Northfield Capital Corporation. Rob is a generalist, involved in several Canadian companies including Southern Star Resources, which merged with partner Exall Resources to form Gold Eagle Mines at Red Lake. He made this endeavour possible without thought of personal gain, out of an interest in mining and a conviction that the story of exploration in this great resource industry should be more widely known among the general public. With the enthusiasm of John Harvey and the backing of Robert Cudney, it was inevitable that the project would come under the aegis of the Prospectors and Developers Association of Canada. For 75 years the PDAC has been a focal point for all aspects of the mining industry. Its influence as the world leader in the field of mineral exploration and consequent mine development has grown steadily.

John Harvey
The Northern Miner

This book will promote interest in and knowledge of Canadian mining for Mining Matters, the education arm of the Prospectors' and Developers' Association of Canada, and give support for those entering the field as a career. Those in the industry at all levels will be reminded of the tremendous benefit the fruits of their labour give to Canada and her people.

Robert Cudney
FNX Mining

Long-time mining man Lloyd Otto summed up the spirit of the mine finder, whether he or she is a prospector, using the traditional pick and a hunch, or a scientist employing the latest exploration techniques. A couple of lines from his poem say it all:

> *So you want to prove you've got guts and spine?*
> *Then go to the bush and discover a mine.*

Perhaps Albert Szent-Gyorgy had prospectors in mind when he observed that a discovery is said to be an accident meeting a prepared mind.

This work needs a disclaimer. The Canadian mining scene is huge and my pen is small. There simply is not enough room in a one-volume account of mineral exploration over sixty years to include every discovery and prospect. The story must, of necessity, tell a little of the great successes in mine finding and a few of the promising developments in the field. Think of this work as a snapshot in time of the metals, diamonds and uranium without the advantages enjoyed by Hoffman in his pioneering work.

He met all of the people he wrote about. The industry has become so huge that in the twenty-first century one would need a personal jet and a huge bankroll to cover the country and meet all those folks at all levels of exploration. Over the past 25 years I have visited many mines, mostly in central Canada. I have explored Eldorado, the first gold mine in Ontario; travelled down the longest mining ramp in the world at Kidd Creek; seen the riches of the combined Red Lake mines; enjoyed chats with people at the Round Up Conference in Vancouver and the Prospectors and Developers Association Conference in Toronto. Beyond that, travel for this project has been limited. Accounts here of the mineral industry in Canada only relate to the metals, diamonds and uranium. Take a look at the Acknowledgements for the names of those who assisted in fact finding for this book; their experience in the industry is widespread. I sincerely hope readers will recognize the author's respect for all those who search for minerals in Canada.

Introduction

Most miners like their jobs, and wouldn't trade places with anybody. There is a real sense of accomplishment at the the end of the working day … that nobody can deny. When my partner and I are working together at the face, there is hardly a word spoken. We each know the job so well and anticipate what has to be done … I think that's a beautiful thing to witness, men working in harmony.

Ron Jessulet, Bathurst miner, 2001, quoted in *Gesner's Dream*

Canada is one of the most productive areas on earth for the discovery of economic mineral deposits. Its large land mass covers the entire spectrum of geological formations. These have been laid down and formed over the past four billion years. Mineral deposits run the gamut of time from early Precambrian to the late Cretaceous period. Low population density, accessibility, generally habitable environment and glacial outcrops help make favourable conditions for exploration. Man has supplemented the bounty of nature's contribution. Canada is fortunate to have individuals who have developed an infrastructure of financial capacity, educational facilities and scientific expertise that is a rich mix of human expertise and resources equal to or better than anywhere else in the world. These people have created Canada's wealth through the careful exploitation of her mineral resources.

Local resident at Golden Bear Mine, Telegraph Creek, B.C.
MABC

Prior to the beginning of the twentieth century, large-scale, economic, producing mines were largely nonexistent in Canada. Any prospecting that was undertaken was carried out by native groups, early settlers and fur traders. These early searchers for minerals were successful mainly by chance. There are many instances where pure luck has found a mine. Gradually such hit and miss exploration was replaced by the advent of newcomers — single-minded, strong-willed

European immigrants — who began to fan out across the country. These men were attracted to the outdoor life and the freedom that prospecting offered, and they made use of emerging knowledge of logically directed, systematic mineral exploration. When successful, their efforts were rewarded by the Dominion government as well as by some private companies. Such recognition did not come from stock markets and financial institutions because these were non-existent in pioneering times.

The earlier nineteenth century explorers paved the way for the era which would follow after 1900. A growing market for resources around the world accelerated demand, first for Canada's vast timber resources, and then for her abundant mineral wealth. Gold was a glamour metal and became a touchstone for an explosion in mineral exploration. The lustrous yellow metal was aggressively sought after the price per ounce rose to $20 U.S. early in the new century. Silver, copper and nickel mines also came into prominence, mainly through the carving of new railways across the continent, but gold was still the prize that spurred prospectors. When President Franklin Roosevelt fixed the price of gold at $35 U.S. in 1933 during the lost years of the great Depression, it was a blessing to Canadian prospectors and miners.

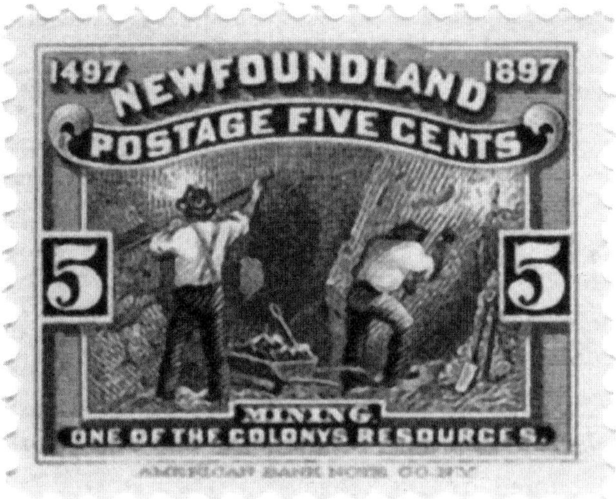

Newfoundland was one of the first places in the world to represent mining on a stamp.
Charles Dearin

This forward-thinking move was supplemented during World War II by a huge demand for strategic metals. During the war, and for close to thirty years, the Canadian government further strengthened the industry by subsidizing gold mines with the Emergency Gold Mining Assistance Act or EGMA. Demand drove the price of gold higher until, in 1972, President Richard Nixon allowed the price of gold to float unfettered and to be governed by market demand. By January 2008 gold commanded above $860 U.S. an ounce and exploration for the metal increased dramatically. Exploration methods until the end of World War II were still primitive compared to those of today. Exploration below the surface was still only attempted in conjunction with the delineation of surface discoveries.

Before World War II mineral production in Canada came mainly from Ontario and Quebec, due to the highly productive, early Precambrian Abitibi greenstone belt. Through its east-west length of 500 kilometres, this metal-rich zone has supplied the bulk of Canada's economic minerals, including no less than 150 million troy ounces of gold. Since then, over the past sixty years, the nation's mineral exploration efforts have resulted in the development of many new mineral deposits throughout Canada. There has been new and continual mining of earlier discoveries containing gold, silver, copper and nickel, as well as new production from zinc, lead, uranium, potash, iron, antimony, tungsten, diamonds and the platinum group metals. There have been other lesser by-product commodities and the new base-metal-dominant deposits have taken over as the largest revenue producers, augmented by income from precious metals.

Relative proportions of Canadian mineral production by deposit type

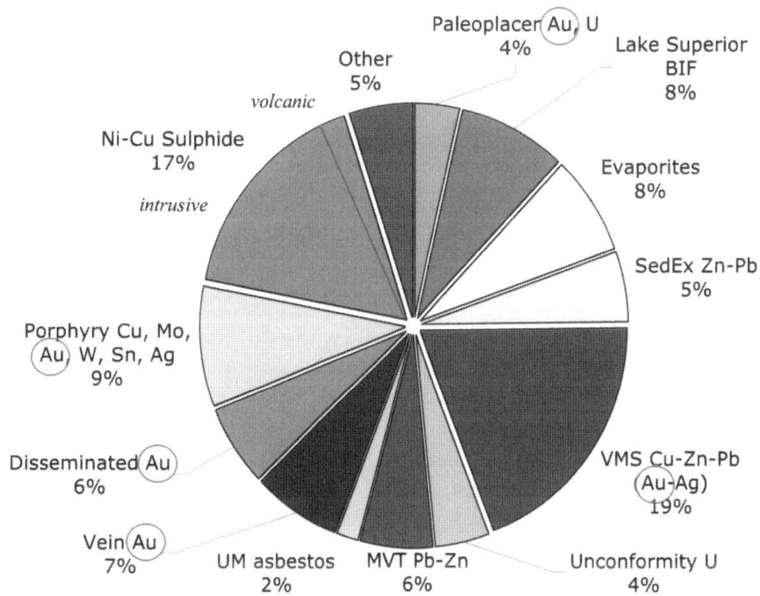

Some insight into several contributing factors is necessary to document the successes of the past sixty years of exploration for hard rock minerals in Canada. So it is that six chapters of this volume explore evolving trends in exploration. Created mainly by Canadians, these trends have placed the country in the forefront of successful explorationists and developers throughout the world. Evidence of this is seen in the large number of both junior and senior mining companies headquartered in Canada and the huge sums they expend. Most of Canada's mineral production is exported and the contribution of mineral production to the country's gross domestic product (GDP) has raised our standard of living to a lofty position among developed countries.

Canada is in a most promising position for the future, providing our mining industry is not crippled by events and trends which have stifled mineral production in less fortunate countries. These include adverse cost increases, development barriers, and environmental malpractices. The above-mentioned six chapters explore movements in mineral exploration, covering the successful methods developed, largely by Canadians, since the end of World War II. These are: exploration funding; geological modelling; geophysical methods; geochemical assessment; improvements in auxiliary equipment; and satellite imaging, which is just beginning to come into its own. While those in the industry apply these techniques in the hunt for mineral discoveries, there is a source of initial stimulus and early development in the university academic programs and government field surveys, which provide supporting research. Other institutions, such as economic think tanks, document objectives and record performance.

Tom Schroeter, Senior B.C. Mines Geologist, in the Toodoggone area

More Than Free Gold is dedicated to the explorationists and their efforts over the past six decades, but it only covers a narrow spectrum of the hard rock industry and its producing mines during this period. The interest here in this work is in the metals, uranium and diamonds. No one volume could cover all exploration efforts including those which did not warrant production. Instead, the book seeks to cover the major mining camps and smaller mines for the period since 1946 and document the importance of Canada's mines to the total GNP of the country. It is hoped that other writers will come forward and record the contributions of all mineral discoveries across the country which have been omitted from this book, and thus give them recognition as well.

There is another purpose to *More Than Free Gold* besides the documentation of mining trends and new mine production over the past sixty years. The first forty of those years — 1946 to 1986 — was a period of an enormous increase of new mines coming on stream, coupled with the development of new exploration technology. The last twenty years, however, has experienced a downturn in this good news story. There has been a decrease in the development of new mines, especially in the base metals area. This has caused potential shortages in concentrates and raw feed from Canada for our smelters. Many mines have closed. The responsibility for this 'black out' period in Canadian exploration may be laid at the door of huge decreases in government funding for supportive field mapping programs and severe downsizing by the major exploration players. Those in the exploration field of the industry have survived this black out period through tax-related government flow-through incentives. A bright light on the horizon is the recent advancement in mineral commodity prices which, if sustained, will offer Canada a chance to grow its mining industry once again. Declining mine revenues may then be reversed and the north country restored to a level of productivity that will support further growth.

As the book draws to a close, two chapters draw parallels to the events affecting exploration funding. These include world economic cycles, the booms and busts, and other demands on mineral production, such as the growing demand for nuclear fuel and explosive Third World economic growth. Exploration budgets have been adversely affected by consolidation and down-sizing by the large mining companies, major exploration scandals and new security regulations. Strong commodity prices have provided the economic success needed to give Canada the impetus and the capital for the discovery of new ore reserves.

This work offers a graphic illustration of the contribution of Canada's hard rock mining industry to the gross national product (GNP) since World War II. The increase in demand for space-age metals, nuclear fuel and other industrial requirements maps the way for the future of

Low impact drilling on a mountain top in Peru by a Canadian crew
PDAC

Canada's mineral resource industries. This will happen if the need is widely understood and the country is ready and able to meet the demand with technology suited to the new millennium.

Geography, distance, population, and other factors clearly indicate that Canada is unlikely to soon become a world-scale industrial power, but it can play a major role for the larger, more heavily-populated centres of the world. This country can supply raw materials in abundance, in a cost-efficient, environmentally-benign fashion that will meet the demands of a fast-growing world population.

Any the support and encouragement that Canadians in all walks of life can give to mineral exploration and the mining industry as a whole will be returned manyfold in lasting benefits to our widespread people.

We Lead

The Prospectors and Developers Association of Canada

Times were tough for Canada in the early thirties. The worldwide Great Depression struck at the Canadian way of life and the welfare of its citizens with devastating force. Millions lost their jobs and industry was decimated by bankruptcies and a lack of demand for products and commodities. Amid all the uncertainty of those lost years, one area of the mining industry prospered. Gold was a commodity that was in high demand and camps such as Val d'Or, Geraldton, Kirkland Lake and Timmins, kept miners employed when many other ventures closed down. Mining people are fighters and, while other segments of

The original founders of the PDAC
PDAC

the economy hunkered down to weather out the storm, thirteen very determined men set out to defend their way of life from what was generally seen as an act of government folly.

In this brief account of what has become a vital national institution, there is a need to recall those individuals who met to protect their industry from ill-founded legislative interference. The original founders were Walter Segsworth, who was a driving force in getting the group together and acted as secretary for the first ten years, Fred Bidgood, Peter Graham, T.H.Rae, Bill Fairburn, Arthur Cockeram, Robert Cockeram, Karl Springer, J.A.Lindsay, Ed Cochenour, Bill Cochenour, Ed Rutherford and Fred MacLeod. They represented a fair cross section of mining endeavours of the time. Engineers were allied with prospectors. The majority were men who had been responsible for finding at least one significant ore body. Some were professionals who had brought properties to production and one had even delineated the great Horne ore body which became Noranda. Among them were both a great promoter as well as a money man.

Prospectors stake a claim at the Royal York Hotel, Toronto, 1968; Hugh Carlson, A. Parres, Don Buckholtz and J.P. Critton. *PDAC*

The issue that galvanized these men was a move of the Ontario government which, in effect, would have relegated prospectors to a back seat in the business. The so-called Engineers Bill proposed that only mining engineers could report on the merits of a property. The government of the day did not understand that at the current stage of the industry, and for many years to come, prospectors would continue to be the prime movers in the hunt for minerals. Few could afford the services of an engineer to evaluate and report on their claims. Only after World War II did the sciences assume a more prominent role in mine finding. In the thirties it was vital that prospectors avoid the imposition of any legislation which impeded their work.

The offending legislation of 1932 was dropped in the face of informed and persistent opposition, and the growing group of mining people who had joined together in common cause finally settled on a name — the Ontario Prospectors Association. The motto adopted for the fledgling group was 'We Lead,' and within a year its purpose was outlined. The association would work to foster the mining business. Prospectors, claim owners, financiers, developers now had a collective voice. The group carefully scrutinized legislation which might threaten the rights of its members. From the earliest days the association became the leading advocacy group for the mineral industry in the country. Politicians of all stripes learned to pay careful attention to the positions held by the association because the leadership made it plain that the good of the nation came first and any initiatives they proposed would be of eventual benefit to all Canadians.

There is a book available in libraries across the country which gives an account of the PDA's first half century: *The Discoverers* (1982, Pitt Publishing). This work gives the background to the association and many of the people who brought its aims forward: the members, staff and volunteers. One person stands out among them all, and that is Viola MacMillan who served the longest term as president, from 1944 to 1964. On the occasion of the silver anniversary of the association she stated: "I am a miner. I love this business, and I want to stay in it until I die." At the 75th anniversary in 2007, the film *Let Me Call You Sweetheart* was screened, enabling members to recall the achievements of this remarkable leader.

Viola MacMillan and Franc Jobin
Museum of Nature, Ottawa

Viola's husband George had preceded her in the president's chair from 1941 to 1943, but in his wife the association found a woman who, though less than five feet tall, was a human dynamo. Both Viola and George had travelled across the country in both hard times and good times in search of minerals. The couple was a rarity as few women were involved in prospecting. Her autobiography, *From the Ground Up* (ECW Press, 2001), documents her hardscrabble upbringing on a small, rocky Muskoka farm, and her need to work to support her family. She was never able to finish high school. Known affectionately as the Queen Bee of Canadian mining due to her boundless enthusiasm and interest in all aspects of the industry, Viola was involved in the discovery of several producing mines. In her role as secretary of the association, she successfully lobbied for exploration incentives for prospectors searching for strategic minerals during World War II and instituted the custom of a full program of lectures and events for annual meetings.

In the years after Viola MacMillan's tenure, the association grew in membership, in its influence in the industry and in the interest it attracted across the country through the efforts of hard-working presidents, directors and volunteers. Commitment of members was important; it must be remembered that for many years there was no paid staff and members helped out in their areas of interest and expertise. Bill Dennis was a strong president who focused on rebuilding the association and pushed for more corporate memberships as part of an effort to build up fiscal reserves. It was during his tenure in 1965-66 that the association finally secured a permanent location in downtown Toronto. In 1978 the association established the Bill Dennis Award for Prospector of the Year. The inaugural award was made by then-president Ed Thompson, who also proposed and worked on *The Discoverers* — a book on the first half-century.

The association's successes of the following quarter century might be mirrored in a few of the achievements of the presidents during that period. Douglas Hume worked for a better tax and regulatory environment, while Lionel Kilburn lobbied for funding for geological surveys. John Larche moved to secure flow-through share financing, and Bob Ginn established the past presidents' advisory council and helped gain support for the Canadian Mining Hall of Fame. Fenton Scott resolved to promote the mining industry and issued an analysis of flow-through share funding, and his successor John Hansuld instituted a strategic review and planning process, coupled with the significant decision to take the convention abroad and market what the PDAC had to offer internationally. John Heslop had as his priority the restoration of regulatory efficiency in Canada to enable the jurisdiction to be competitive. Gerald Harper's primary vision was to build stronger relationships with other industry groups to maintain a united front and promote advocacy on a full-time basis. He was followed by John Steele who spearheaded a successful campaign to have the federal government establish what became known as the investment tax credit for exploration, (the 'super flow-through' program). Bill Mercer had as his chief goal the development of a new strategic plan, and he also moved to have the association become a leading information-sharing forum for exploration communities via the medium of the annual convention. Long-time industry volunteer Peter Dimmell made it his business during his term of office to meet with as many prospectors as possible across the country to discuss their concerns. Among the priorities of Patricia Dillon, the president during the 75th anniversary year, was a successful push to extend the super flow-through program and to reinforce the partnership between Mining Matters and the Canadian Institute of Mining, Metallurgy and Petroleum in the staging of the Mining in Society show.

PDAC president Ed Thompson presents award to Mrs. Mercer, 1978.

PDAC

The PDAC was the first association to organize an annual convention specifically for the exploration sector. The convention has become the chief focus of the association in attracting both national and international attention to the mining industry. Executive Director Tony Andrews, working with volunteers and staff, moved in 1992 to have the association take the convention international. As a result attendance at the annual March event has ballooned to a record high, more than 17,600 in 2007 — the diamond anniversary year. Since the function brings in 75-80% of revenues, it has become a secure base to fund the wide spectrum of programs that members now enjoy.

Peter Ferderber, Prospector of the Year, 1980

PDAC

No federal government has been able to ignore or confess to lack of knowledge of Canadian mining achievements and concerns since the inception of the association in 1932. Concerted and continual representations by the association to the legislators have ensured recognition, respect and an awareness of this vital industry. In 2007 the annual PDAC brief to Canada's mines ministers, formulated by the PDAC after consultation with all other mineral industry associations in Canada, set the scene by crisply pointing out that there had been a severe downturn in the industry between 1997 and 2002, that base metals inventories were declining and few new discoveries had

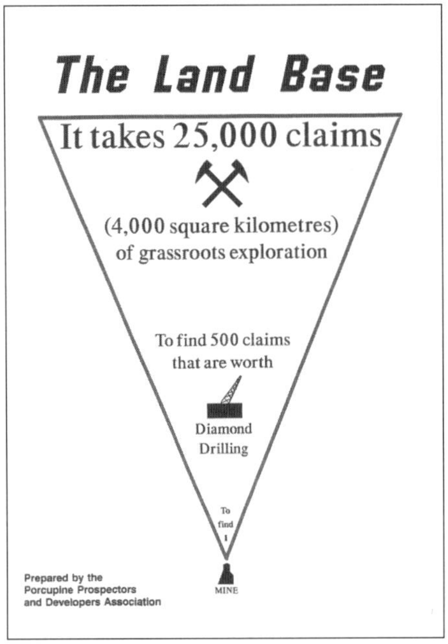

been made. Reserves had to be replenished and strong investment in exploration must be a priority. One important remedy suggested was the continued support for flow-through initiatives and a strengthening of Canadian exploration expense allowances. The Canadian Mineral Exploration Tax Credit, and its predecessors, the Investment Tax Credit for Exploration and the Mineral Exploration Depletion Allowance, have been very successful programs for mineral exploration in Canada. They have contributed both to significant mineral exploration in the country and new mineral discoveries. Funds accruing from the program have kept exploration dollars in Canada and have ensured Canadian competitiveness in retaining and enhancing exploration investment. Another thrust of the brief was the request that long-promised funding for cooperative geological mapping strategies be honoured.

A recurring theme in all of the presentations has been the need for long-term strategies. A mine can take many years to discover; then it takes time to define an ore body, and to plan, permit, finance and build an infrastructure. The PDAC takes the position that the industry would find it easier to raise funds if there were a national securities regulatory body, but so far this concept, well-accepted in other countries, has not come into being. Alberta has joined Ontario and the federal government in support of a single body.

Dit Holt and Norman Keevil, Jr. start a hockey game, 1980.
PDAC

A major achievement of the association is the creation in 1999 of a scientifically-based approach to the protection of the environment and a carefully thought-out process of community engagement which is so all-encompassing that it is being adopted by progressive mining concerns around the world. A volunteer group of mining, exploration and environmental professionals worked to produce a manual for the industry which became known by its initials, ***e3***, for Environmental Excellence in Exploration. This free manual is so thorough in its approach that the table of contents alone now runs to sixteen pages on the website. Buttressing the document is an Exploration Code of Conduct. There are sections on management essentials and technical content. The checklists are, in effect, environmental compliance forms. Actual case studies are provided as examples for those using the documents. Proof of the acceptance of Canadian leadership in ***e3*** comes in the fact that more than 1,700 users in forty countries are registered as following its principles and the materials are now offered in French, Spanish and Portuguese. The PDAC has its own manager for ongoing progress of this environmental initiative and has an annual award for Environmental Excellence in Exploration to keep the movement front and centre worldwide.

Lloyd Rochester, a prospector, pilot and mining engineer, discovered four major deposits and was the founder of Prospector Airways.
PDAC

One of the most exciting education projects of the PDAC may be interpreted as a gentle pun. Mining Matters indicates that mining *matters* and partners with this truism to offer information about the industry. This education arm of the association presents schools with opportunities for students to learn about the importance of rocks, minerals, metals and mining, and to make the connection between these resources and our everyday lives. Wisely, the programs are developed in conjunction with science teachers and, following provincial curriculum guidelines, focus on grade four, where youngsters have mastered reading, and on grade seven, a level where their curiosity about our world is rapidly advancing. Complementing the project at the high school level is Discovering Diamonds, a course which offers twenty-one activities relating to this valuable resource. The materials and direct expenses of taking courses to the schools are funded from the donations to the popular Canadian-registered charity. Many mining companies support PDAC Mining Matters, as do individual members of the industry and public. There is an added bonus for the future of Canada's mining industry: some students may follow up on their initial introduction to resources by pursuing studies in the earth sciences.

Even the Royal York waiter got into the act, 1982.
PDAC

For many the public face of the mining industry and the association which works on its behalf is the annual International Convention, Trade Show and Investor's Exchange. The PDAC has 7,000 members within Canada, and many more travel from far-flung parts of the world to attend the event, often meeting up with friends and former colleagues to renew old acquaintances after a year working on remote mines or ventures. They catch up on gossip, prospects and even swap job opportunities. Canadian delegates reflect a fine balance here between urban and rural areas, with a large contingent from the North. Among them are the prospectors. New Brunswick prospector Dave Stevens echoes the popular attitude: "If you catch or find something, it's a great day. If you don't, it's still a great day. Money is part of it but that's not your prime focus. You have to love and respect the woods and the outdoors."

Prospectors Walter Baker and Don McKinnon, 1984
PDAC

There are more than 450 exhibitors displaying the latest technology, products and services, and delegates from over 100 countries attend the event. The Investors Exchange connects the investing public with junior mining companies. In this popular convention segment, the PDAC expands its technical program to include an investors' forum. Since 1994 mineral exploration companies have gathered at the Exchange to promote their prospects and meet up with people who take shares in their ventures.

None but the inveterate attendee can say that they will experience all of what the convention has to offer. Throughout there is a great mix of interests and commitments. Prospectors, geologists, engineers, academics, students, developers, financiers, bankers, analysts and government people exchange views and convictions. Networking is expected and delegates have opportunities to interact and conduct business with colleagues sharing mineral exploration interests from all parts of the world. Industry students find summer jobs and connect with prospective employers. Deals are made and agreements are signed in company hotel rooms which, over time, create jobs and wealth for Canadians. Few realize that eleven months of planning by committees of volunteers have gone into the development of a vast choice of experiences for delegates. The Metro Convention Centre, the largest such venue downtown Toronto can offer, is booked solid for this much-anticipated gathering of all things mining.

Kirkland Lake Mayor Joe Mavrinac worked with the 'Go with the Flow' movement to preserve government support for the mining industry, 1987.
PDAC

The 75th Anniversary convention was well-celebrated with the largest number of registrants ever, from all facets of the industry worldwide, in attendance. There was a calendar of mining cartoons for delegates and a comprehensive anniversary publication. People are accustomed to picking up pins at various booths, but the PDAC awards night dinner gift topped these souvenirs. For years to come those who attended that celebration could sport a very special rock hammer pin. The memento was sterling silver and gold-plated with a tiny diamond as a nod to this milestone celebration of the largest annual mining function in the world.

John and Delores Larche flank Viola MacMillan at the 1987 convention.
PDAC

There can be no doubt that the convention will remain the central mineral exploration event for years to come. The global demand for commodities will boost attendance. Canadian prospectors, developers and junior companies will continue to play a worldwide role in opening up mineral deposits. Canadian financial markets will continue to be the place to come to raise capital for exploration projects around the globe.

The diversity of the elements at the convention reminds one of the outreach and involvement of the PDAC in mining affairs both across Canada and on the international scene. The association board has 48 members representing all regions of the country and every sector of the industry. Representatives of the PDAC liaise and work with sister organizations that make up the Canadian Mineral Industry Federation.

1994 Yukon convention members show off a back-home road sign which in name could be compared with some less fortunate prospectors. *PDAC*

Beyond the regional industry groups, the association has close ties to other land use stakeholders and resource management interests. Experience in this pan-Canadian approach naturally leads to interaction in the world arena. PDAC staff and board members attend and often give presentations to conferences and meetings in many countries and, when requested, offer advice to officials in other jurisdictions. Through participation in events such as the World Mines Ministers Forum, closer international ties are forged. The PDAC is also an associate member of the International Council on Mining and Metals, which is a key contributor to the sustainable development movement.

Convention scene at the 2007 show with 17,600 registrants.
PDAC

Increased revenues generated by the convention enable the association to conduct advocacy that is strongly supported by members. Much of the work is done by volunteer committees. One initiative of the organization has been a move to improve mining standards and regulations for the benefit of investors and for mining and exploration companies. This work resulted in National Instrument 43-101, now recognized as the global standard for mining disclosure. A major outreach of the association is to defend the industry's right to land access. The well-received Land Use Strategy works to limit acreage removed from exploration, and respects the interests of the variety of stakeholders competing for land access by promoting integrated landscape management. In the area of Aboriginal affairs, prime moves by the association include support of resource revenue-sharing and close monitoring of land claims. Corporate social responsibility is a prominent advocacy area which recognizes a duty of care to all stakeholders within the industry, customers and local communities. This thrust addresses social issues, the environment and sustainable development. Among finance and taxation initiatives, the PDAC lobbies for tax incentives to increase base metal exploration and to further utilize known mineral areas by obtaining tax credits for deep drilling. One aspect of the work of the Geoscience Committee is to ensure that those industry professionals working in multiple jurisdictions can do so without incurring multiple registration fees. With an eye to future recruitment for the industry, the PDAC partnered with sponsors to provide a mineral exploration workshop in Sudbury for post-secondary students in May, 2007.

Meet the Miners session at the Ontario legislature: P. Bousquet, ***e3***; Laura Clinton, Mining Matters; and E. Freeman, geologist.
PDAC

New programs are also funded by revenues generated through the convention. Among these is a program focusing on the need for protection of worker health and the provision of a safe working environment. Guidelines for the industry are being developed and courses in first aid will be set up. Another welcome innovation has been an arrangement with the commercial insurance industry whereby independent geological consultants may obtain a wide range of coverage for risks and liabilities.

The PDAC mission statement declares that the association exists to protect and promote the interests of the Canadian mineral exploration sector and to ensure a robust mining industry in Canada. The PDAC began in hard times with a conviction that the protection and advancement of the mining industry go hand in hand with a strong and prosperous Canada. That conviction holds true as much today as it did more than 75 years ago; Canada has become a global mining powerhouse.

Paying for It All

Trends in Exploration Funding

Some producing mines are proving up
But Prospectors are not too bright;
For as time has passed, the efforts wane
Though there's many a muskeg underlain
With untold riches in hidden vein
Which drilling may bring to light.
But a Prospector only has time to spend
Who else will share the cost?
The camp is BIG; the formation right;
There must be a hundred mines in sight
But they can't be developed overnight
It takes money and lots of time.

A.L. Reading, *The New Road to Red Lake*

Deep pockets or ready access to large amounts of someone else's cash are needed to survive in the business of mining. The Metals Economics Group (MEG) of Halifax, Nova Scotia is a world leader in mining industry intelligence. The MEG has sounded an alarm, stating that between 1980 and the present there has been a significant decline in proven and probable Canadian mineral reserves. During the period from 1980 to 2004, all the major base metals have seen a fall in production. For example, in mid 2006 it was estimated that Canada had only 5.5 years of lead reserves, 7 years of zinc, 10.5 years of copper and 21 years of nickel. At the same time, exploration levels are higher than they have been in many years. Unfortunately, mine finding has not been consistent over time. At present there is an extended boom in the commodity price cycle. But, will investment in exploration grow during this period of prosperity?

Thayer Lindsley has been called by some the greatest mine finder of all time.
Falconbridge

Although many investors have switched to less risky assets, there are now many wealthy entrepreneurs and venture capital groups who have been attracted by the surge in metals prices. This is an encouraging trend and should facilitate exploration financing. The late Walter Gordon, Minister of Finance, had reason to speak positively of mining in 1964: "As the major single source of foreign exchange, the mining industry is the foundation of the modern Canadian economy ... it is unquestionable that the mining industry will for long continue to play a major role in contributing to the real wealth of our country." Year-end production figures of metal mines in 1964 showed that over the previous six decades dividends paid by the industry had totaled over four billion dollars.

Large mining companies were the principal source of exploration funding from 1946 to 1980. After this time, due to various factors connected with their bottom line profitability, they began to dramatically decrease expenditures aimed at locating new ore bodies. Fluctuations in market prices as a result of changes in world economic conditions and global economic events have always affected mineral and metal prices. The major oil companies took up some of the slack in mineral exploration by investing heavily in the field from about 1970 until the late 80s, when budget cuts in the oil patch due to low per barrel prices rapidly dried up funds for company prospecting. In recent years, industry consolidation and major company acquisitions have actually resulted in a decline in exploration budgets. Fifty-five major mining and exploration companies worldwide were swallowed up in mergers and acquisitions in the period from 1997 to 2004. This cut about $280 million U.S. in exploration budgets.

Between 1946 and 1990 capital provided via the brokerage community and allied promoters was a huge source of exploration funding. Latham Burns, former head of Toronto brokerage company Burns Fry, has stated that there is now very little financing of exploration ventures undertaken by companies independent of the mining industry giants. There has been a shakeup in the brokerage community and with internet trading now prominent there are far fewer traditional personal service brokers. In 2005 the Canadian Venture Capital and Private Equity Association reported a decrease in new money for investment with a new emphasis towards the life sciences sector. Since 1990 banks have taken up the funding role formerly provided by brokerage houses.

Banks favour sure things. They have come into the mining business because much of the risk of mineral exploration and development has been eliminated due to the thorough way in which modern exploration outlines and proves ore bodies. Debt financing in mining is still easier when the banks are offered opportunities in major developments involving large sums of money.

Major companies have assets to back the debt they assume. Banks hold the whip hand in financing as they always insist on a preferred security position over the position of company shareholders. These canny institutions have their own research departments which produce studies on the economics of a proposed operation so that an evaluation of risk is extremely accurate. The growth of the uranium industry in the mid-1950s was due to bank financing, no doubt after a strong push from the federal government. Bank financing of exploration and development is still available but the process is slow.

A promising innovation in mine financing is the pairing of independent engineering and financial advisors to assess lending risk. Recently, Hatch Consulting and Rosco-Postle examined the risks in developing the world's deepest single-lift mine shaft for the largest underground gold mine in Canada. In another project, Hatch assisted in reviewing a major mill expansion and mineral processing plan along with the financial models, contracts and environmental planning associated with it. Roscoe-Postle assessed the reserves, geology and mining. Hatch then assisted with the structure of the loan agreement and monitoring of cash flow and environmental procedures. In this manner the lending institution was able to clearly evaluate and minimize the risks of the venture.

Ask any group of mining people about the biggest problem plaguing the mineral resource industry and the answer is always unanimous — "too much government." High taxation and the inclination of the provinces to take a larger share of royalties have put a damper on the once-favourable atmosphere which encouraged high-risk investment for a good return. In 2007 the Liberal government of Ontario put a royalty tax on diamonds without any consultation or warning for De Beers, the company involved. Since the early eighties government has generally come to realize that support is vital to nurture the Canadian mining industry, the goose that lays the golden eggs. The leader in the fight for significant tax incentives has largely been the regulatory affairs section of the Prospectors and Developers Association of Canada.

In the late seventies the federal government finally acknowledged that the message about exploration funding was getting through. A program was instituted permitting the use of what are commonly known as 'flow-through shares.' This meant that junior mining companies could provide an exceptional incentive to the investor. In return for risk capital, investors in the highest marginal tax rate received a flow-through of a tax credit equal to 100% of their investment. The plan saw a more competitive edge for Canada in the exploration field and kept junior companies not only alive but able to offer an attractive investment opportunity to the public.

This sketch, if anything, flattered Murray Pezim.

This tax incentive helped the industry in the down cycles of mining activity for two decades, but it was not until 2000 with the announcement of the 'super' flow-through shares or, more formally, the Mineral Exploration Tax Credit, that the reinforced opportunities for mining financing gained momentum. PDAC President Patricia Dillon estimates that from 2000 to 2006, the exploration tax credit had generated $1.4 billion in exploration spending, and most of this was spent in areas needing employment — Canada's northern and rural regions. The results averaged out to one significant discovery a month from the commencement of the program in October 2000. But even with this impressive record, the much-hoped-for legislation was not supported by the outgoing Liberal government and expired at the close of 2005. The result was a concerted effort by the PDAC and others to have the 'super' flow-through tax break extended.

The lobbying effort paid off and the tax relief venture was extended for two years as confirmation of an election promise by the Conservatives. The PDAC brief pressed for continuation of the Investment Tax Credit and for a number of other initiatives. The brief successfully proposed that the Canadian Exploration Expense should include the costs of community consultation, environmental studies and feasibility studies; that tried and true former base metals mines locations should be certified as exploration expense; that credit should be given for deep drilling costs; and that funding should be committed to a geological mapping strategy. The PDAC followed up the success of their lobbying effort by continuing to press all levels of government to implement a long-term exploration strategy, including accelerated processing of Aboriginal land claims, the training of new workers and a renewed push to have a single securities regulator for the whole country.

The 'super' flow-thorough was made even more attractive to investors for it offered an additional 15% federal tax credit for grassroots exploration. British Columbia (B.C.), Manitoba and Quebec also provided additional tax incentives. The Mineral Economics Group indicates that Canada has been the number one country in the world for mineral exploration since 2002. In 2005 close to one fifth of total world exploration expenditures was for projects in Canada. The pay-off in these tax credit incentives has been seen in grassroots discoveries by junior mining companies of gold at Hemlo, diamonds at Lac de Gras, and the huge nickel and copper find at Voisey's Bay.

Much of the glamour of the mining business has not come from the major firms in the industry and certainly not the rather slow-moving and staid banks. The excitement and a large proportion of the exploration success in Canada have been generated by the junior mining companies and the promoters who have raised their funds. Junior miners usually start their business careers with a few claims of rocky ground and a dream. There are many in the field. In 2006 there were

Ontario Mining Operations

GOLD MINES

	Name	Company
1	Clavos Mine	St. Andrew Goldfields Ltd.
2	David Bell Mine	Teck Cominco Limited, Barrick Gold Corporation
3	Dome Mine	Porcupine Joint Venture – Goldcorp Inc., Kinross Gold Corporation
4	Eagle River Mine	Wesdome Gold Mines Ltd.
5	Hoyle Pond Mine	Porcupine Joint Venture – Goldcorp Inc., Kinross Gold Corporation
6	Macassa Mine	Kirkland Lake Gold Corporation
7	Musselwhite Mine	Goldcorp Inc., Kinross Gold Corporation
8	Pamour Mine	Porcupine Joint Venture – Goldcorp Inc., Kinross Gold Corporation
9	Red Lake Mine Complex	Goldcorp Inc.
10	Williams Mine	Teck Cominco Limited, Barrick Gold Corporation

BASE METAL MINES

	Name	Company
11	Kidd Creek Mines (copper, zinc)	Xstrata Copper
12	Lockerby Mine	First Nickel Inc.
13	McCreedy West Mine	FNX Mining Company Ltd.
14	Montcalm Mine	Xstrata Nickel
15	Redstone Mine	Liberty Mines Inc.
16	Sudbury Operations: Copper Cliff North Copper Cliff South Creighton Garson Gertrude McCreedy East/Coleman Stobie	CVRD Inco Limited
17	Sudbury Operations: Fraser Onaping/Craig Lindsley	Xstrata Nickel

PLATINUM GROUP METAL MINES

	Name	Company
18	Lac des Iles	North American Palladium Ltd.

OTHER METAL MINES (Magnesium, Calcium, Strontium)

	Name	Company
19	Timminco Metals	Timminco Ltd.

MAJOR INDUSTRIAL MINERAL OPERATIONS

	Name	Company
1	AMP Quarry (carbonatite)	Agricultural Mineral Prospectors Inc.
2	Badgeley Island Quarry (silica)	Unimin Canada Ltd.
3	Blue Mountain Operations (nepheline syenite)	Unimin Canada Ltd.
4	Cavendish Twp Mine (vermiculite)	Regis Resources Inc. - Vermiculite Canada
5	Goderich Brine Field (salt)	Sifto Canada Inc.
6	Goderich Mine (salt)	Sifto Canada Inc.
7	Hagersville Mine (gypsum)	CGC Inc.
8	Henderson Mine (talc)	Canada Talc Division of Dynatec Minerals Division
9	Kapuskasing Phosphate Operations (phosphate)	Agrium Inc.
10	North Williams Mine (barite)	Extender Minerals of Canada Ltd.
11	Ojibway Mine (salt)	The Canadian Salt Company Ltd.
12	Penhorwood Mine (talc)	Rio Tinto Minerals Group
13	Tatlock Quarry (calcium carbonate)	OMYA (Canada) Inc.
14	Windsor Brine Field (salt)	The Canadian Salt Company Ltd.

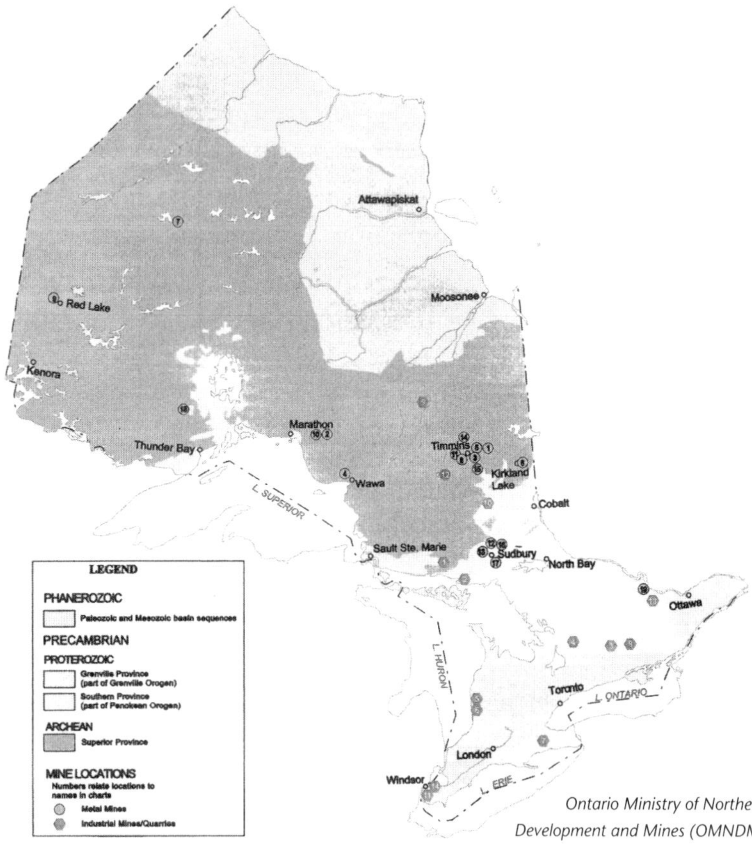

Ontario Ministry of Northern Development and Mines (OMNDM)

1,300 exploration companies and more than half of these had properties in one hundred different countries outside Canada. The current junior-led recovery in exploration relies on their ability to raise capital, sparked by high metals prices and the federal tax incentives. When a junior acquires a very promising discovery, the firm may enter into a joint venture with another company and spread the risk, as well as the promise of reward. Eventually juniors bow to the fact that they do not have the resources to develop a large scale property and so they make a deal with one of the majors.

Since 1950 promoters have generated the buzz and excitement to enable juniors to move their stock. Someone once said, unkindly, that mining stock is to be bought and sold — not loved. Many of the great mining companies that brought joy to the investing public, such as Lake Shore,

Wright-Hargreaves, Canadian Malartic, Little Long Lac and Steep Rock, were made possible by promoters. One promoter who did his share of moving mining certificates was Murray Pezim. He has been reviled and accused of shady dealings and yet he has had great successes and generated enormous wealth for Canada. Pezim entered the field from the brokerage business and parlayed a small junior, International Corona Resources, into prominence by raising funds for a drill program at Hemlo near Marathon on Lake Superior. When the latter partnered with the giant Teck Corporation, the junior company developed the ore body which would become the Williams Mine, for several years the biggest gold mine in Canada. Pezim also played a role in developing other successful mines. I noticed that the promoter kept Mark Twain's words framed in his Vancouver office: "October. This is one of the particularly dangerous months to speculate in stocks. The others are July, January, September, April, February, November, May, March, June, December and August." Murray Pezim once told the author that "The Pez never forced anybody to buy stock," but folks bought; some lost their shirts; and many others made money.

TRIAX

THE GROSSO GROUP

www.grossogroup.com

SNOWDEN

THE NORTHERN MINER

THE GLOBAL MINING NEWSPAPER

Rio Tinto
Memorable
US$7.4-billion year 3

Centerra Gold
Forgettable '06 16

B1-B8

SXR moves to acquire UrAsia

Revised feasibility study buoys Yukon Zinc

Delay at Cigar Lake

MARKET IN LIMBO

Tough year for De Beers

RECORD PRODUCTION, DECLINING PRICES

Queenstake, YGC to tie knot

YGC TO FINANCE NEW COMPANY

Expertise in all aspects of mining.
Contact Wayne Egan at (416) 947-5086 or at wegan@weirfoulds.com

WeirFoulds LLP

The Northern Miner

There are times when the mining industry becomes its own worst enemy. Events that occurred in 1964 and 1990 resulted in the appointment of security commissions and restrictions on the way mining companies could raise funds. The great American statesman of finance Bernard Baruch once described his occupation by stating, "I am a speculator." Speculators large and small are the engine that provides mining funding, but they have to exercise some judgement. Mining is a risky business. One-time president of the Toronto Stock Exchange Howard Graham put it succinctly: "The Exchange cannot protect the speculator from his own stupidity and his own cupidity." A case in point was the Windfall promotion which took place in 1964 that shook public confidence in the mining industry and its regulators to the very core.

Texas Gulf found what would become one of the largest base metals mines in the world in the area of Timmins, Ontario. Veteran prospectors and mining promoters George and Viola MacMillan picked up claims for their appropriately named Windfall Company and the stock rocketed from 31 cents to more than $5 in just over two weeks. Viola manipulated the stock in a process called 'wash trading' in which she bought and sold her own shares, giving the impression of much activity. By delaying reports that showed that the ground was worthless, she made more than a million dollars. After a lengthy trial she was convicted of manipulating the stock and went to prison, although she was released after six weeks of a nine-month sentence. Stricter listing requirements that followed the affair saw an exodus of mining companies and investment to the far less tightly regulated Vancouver Stock Exchange.

Over the next thirty years there were other mining scams including the 1980 New Cinch affair where the Vancouver Exchange added to its dubious reputation when many investors lost their shirts over an alleged gold strike in New Mexico.

Inventive promoters can persuade the public of anything. Robert Friedland, when pushing Diamond Fields Resources, once said that stock promoting was like making a movie. "You've got to have stars, props and a good script." American investment portfolio manager Peter Lynch wrote in the book *One Up on Wall Street* (Penguin Books, 1990) that "a flashy name in a mediocre company attracts investors and gives them a false sense of security." He warned that stocks beginning or ending with 'X' seem to have a fatal attraction for investors. The classic among these was Bre-X, a penny stock that climbed from a few cents in 1988 on the hype of a world-class gold discovery in Borneo to $286.50 CDN by the time it collapsed nine years later. The buzz in the early nineties had built up the find to the largest gold deposit of the century which was said to exceed all the yellow metal that had been taken in the California gold rush. The market value of the stock had run up to $4 billion before the bubble burst when core samples were found by independent evaluation to have been salted. The major losers, other than countless small investors, were major Canadian pensions and other investment funds. The project geologist jumped to his death from a helicopter when the scam was revealed and public confidence in the

mining industry fell to an all-time low.

The Investors Exchange is the Prospectors and Developers Association's annual mining investment show for delegates in Toronto. The location is appropriate as the Toronto Stock Exchange dominates the world as a centre of mining finance. In 2005 there were 1,192 mining companies doing business through the Exchange, far more than the combined number doing business in all other world trading institutions. The value of shares traded was $158.7 billion and the value of mine financings in the city exceeded all other world bourses. Prospectors and the just plain curious can browse 450 booths at the PDAC, attend a forum where companies make promotional presentations, and meet up in the Prospectors Tent with the first explorers on the ground, who proffer maps, samples and drill core to display their holdings. All sources of potential investment are courted by the exhibitors. This show-and-tell is an eye opener for potential investors at all levels and it gives the visitor an insight into the directions in which exploration funds are being channeled. Exploration requires constant drawdown on cash reserves for administrative costs which is, perhaps appropriately, known in the mining industry as 'burn.'

No one in the mining business would argue with the opinion expressed by PDAC board members in March 2002 that the most important issue facing the exploration sector in Canada was the cost of raising capital. Statements like that made by promoter Robert Friedland in 1988 do not inspire confidence. He said in an interview with *Report on Business* magazine: "A start-up company is germinated by smoke and mirrors ... you have to generate the belief that anything is possible." By contrast, the current securities regulatory regime offers many hurdles for mine financiers to overcome that call for reasonable proof of resources. There are multiple jurisdictions and a profusion of securities policies and regulations, all of which force companies to spend hard won funds and precious time raising capital for exploration projects. Junior mining companies account for a quarter of Canada's publicly listed companies. All must undergo the expense of professional advisors in every jurisdiction in which they wish to do business. Anxious to avoid the possibility of scams such as Windfall and Bre-X, the Canadian Securities Administrators set up a rule known as National Instrument 43-101 that governs how issuers disclose scientific and technical information about their mineral projects to the public. Since 2001 the NI 43-101 requires that all disclosure calculated to attract potential investors must be based on the advice of a 'qualified person,' in some cases independent of the project. The standards that govern disclosure in the area of resources and reserves are in line with guidelines approved by the Canadian Institute of Mining and Metallurgy.

The qualified person is an engineer or geoscientist with at least five years of experience in exploration, mine development or operation, or project assessment. This individual should have experience relevant to the subject of the mineral project and the technical report which will accompany it. To cap all of this, the qualified person must be in good standing with a professional organization. The QP has significant responsibilities and also potential liability in areas where professional judgement is required. The impact of NI 43-101 on the mining industry is considerable. When Adriana Resources commenced work in 2006 on

A junior company, Barker Minerals, holds a shareholder appreciation night.
Barker Minerals

The Miners' Memorial, Kirkland Lake, Ontario

Rob Mohr and Sally Lawrence, Museum of Northern History

an iron formation north of Schefferville, Quebec, the company was quick to cover itself by pointing out that its deposits were drilled in the 1970s, thus predating the current disclosure standards. In its 2004 annual report, Western Quebec Mines indicated that it believed that it was in compliance with a host of regulations. These included federal, provincial and local laws governing prospecting, development, production, taxes, labour standards and occupational health, mine safety, toxic substances and environmental protection. The environmental standards are particularly stringent. In addition, a rehabilitation plan for the mine site, a closure plan, and security to cover such costs must be forthcoming. Other provinces and territories have similar requirements.

Problems associated with NI 43-101 include the inability of the operator to disclose information about a mine that was in operation and then decommissioned prior to 2001, most likely due to low commodity prices. The compliance process must still be engaged in instances where such a property is revisited. But, thanks to this nationwide securities regulation, mine salting or core tampering are extremely unlikely to occur. Banks, securities firms and corporate finance types rely on NI 43-101 reports.

The industry would like to see the wealth of regulations relating to mine financing administered by one regulator applying one set of rules in a consistent manner across Canada. Securities laws should provide junior companies with access to capital markets in a timely, efficient and cost-effective process. In the meantime disclosure rules remain such that investor confidence in mining ventures may be maintained.

Exploration funding over the past sixty years has trended towards close regulation of public security offerings and, as a result, a very healthy climate for investment at all levels in the mine financing arena now exists.

Prospecting Off Site

Geological Modelling

Some common ground exists between the approaches of a geologist and that of an historian.

Morris Zaslow, *Reading the Rocks*

One of the most dangerous side effects of geological study for mine finders is that of selective vision, the tendency in people to see what they want to see and not see what they don't want to see due to the psychological impact that seeing or not seeing would inflict on them. In other words, it is essential to stick to the facts. Until the 1950s most ore being mined was being found through surface exposures. Diamond drilling and trenching revealed the character of the rocks beneath the surface, but any other decision regarding an ore body was based on an educated hunch. A valuable set of tools has developed since the mid-twentieth century that has made initial investigation of promising mineral sites much more reliable. The term 'geological modelling' may be said to include methods that allow the geologist to visualize, interpret and present both surface and subsurface data. Such data are presented in the form of high quality graphs, models and plots.

Visitors at a high Arctic camp

Indicator Minerals

Take a look at the phases of exploration, which may be old hat to many, but serve to remind us of the risks involved in mining investment. A property must be assessed to see if the ground has the potential to hold economic minerals. This involves a variety of physical observations, could take many weeks and cost several thousand dollars, and may lead to a success rate of one-in-ten among properties surveyed. If the situation appears promising, the next stage aims to identify a target (which will likely take months), and involves expenditure of tens of thousands of dollars. As geologists know, the rejection rate is, again, 90% of projects undertaken. If the venture survives this far, target testing may take many more months, run into the hundreds of thousands of dollars and still follow the ten per cent pattern of success. As a result, only one venture in a thousand makes a mine. We may step out of the picture at this point but the evaluation phase, mine development and production run into the hundreds of millions.

Those initial steps taken to investigate a promising mineral property will include geological modelling. This will determine whether deposits under review are capable of supplying a quality mineral product on a consistent basis. Surface outcrops, trenching, preliminary diamond drilling, sample assaying and the availability of a good topographical map will enable the collected data to provide a serious picture of the site. The information may be fed into a computer program to provide a preliminary geological resource map. The resulting deposit evaluation should verify a project's potential before significant funds are committed in further steps.

It is a widely-supported premise that innovation in geological modelling must be academically driven, and that fundamental and long-term research must be supported by governments and universities. Research in multidimensional interpretation and visualization of geological data, and the development of new methods and innovative technologies in natural resource evaluation are essential to a healthy mining industry. Mining Association of Canada members cooperate with groups such as CANMET, the energy and technology centre, the main research and development arm of Natural Resources Canada. COREM is a Quebec-based research group that links industry, provincial government and university interests. Another location for research priorities is the Mining Innovation Rehabilitation and Applied Research Organization or MINARCO.

Much has to be done by government in geoscience. In Ontario projects such as Discover Abitibi and the Northern Boreal Initiative, as well as work on geological terrains around Lake Nipigon, are leading to significant mining company interest in these areas. Ontario has also committed substantial funding for geological mapping in its far north. One result has been the De Beers' new diamond mine in the James Bay Lowlands which will open in 2008. The National Geological Surveys Committee is working on mapping strategies across the country as a preliminary survey that should result in a ten-year multi-million dollar commitment which should lead to new resource-based regional economic development opportunities in frontier areas.

Geological modelling is useful in the search in volcanogenic massive sulphides (VMS) for base metals such as lead, copper and zinc. VMS districts occur in Buchans, Newfoundland, and at Bathurst, New Brunswick.

Electromagnetic (EM) and magnetic (MAG) surveys can detect these deposits because they are electrically conductive and magnetic. In the Abitibi district of Quebec known base metals regions are being examined using a flow chart known as an inference model which makes use of geological as well as geochemical and geophysical inputs. An example of government and industry cooperation in geological modelling appears in the Geological Survey of Canada's Flin Flon, Manitoba, project. This five-year effort is taking place on the property of Hal Resources at its Sherridon location.

Geological modelling is particularly useful in exploring for gold in what is known as 'deformation corridors.' A study of structural geology leads to data on three-dimensional distribution of rock bodies and their 'planar' or folded surfaces. Corridors of rock which have been deformed as a result of applied force such as pulling, pushing and twisting of the earth's crust, exhibit much folding. Deposits of metals are historically

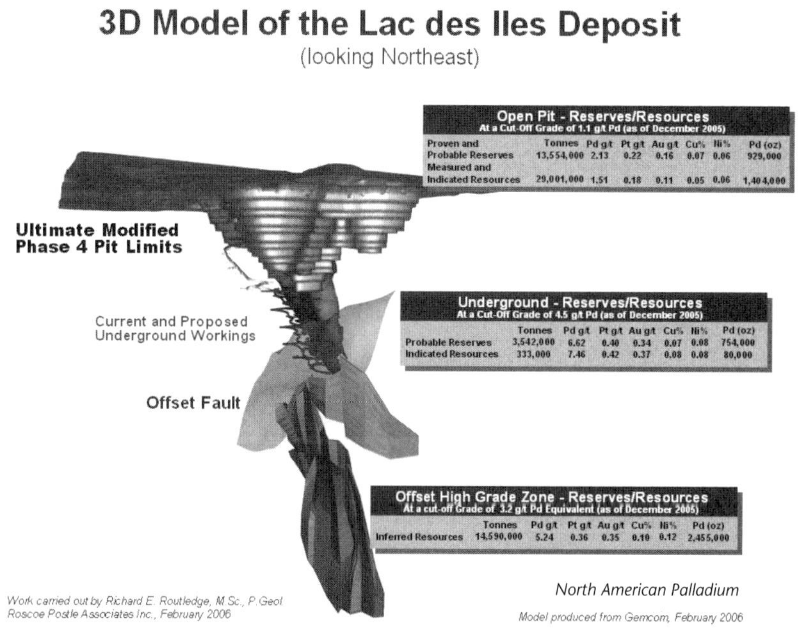

found in these areas. After field work, the input laboratory and computer modelling has been likened by one university researcher to the moral of a story. The results enable one to see through the rationale and use it for guidance in future investigation. Plato Gold is working on a large deformation corridor in the Val-d'Or area of Quebec and also in the Harker-Holloway area of northeastern Ontario. Wolfden Resources' promising Redauraum property at Red Lake is sited on a major corridor. In Nunavut, Miramar and Hope Bay Gold have committed their prospects on a deformation zone at Hope Bay.

Notable unconformity uranium deposits are found in Canada at Elliot Lake, Ontario and in northern Saskatchewan. Deposits in Saskatchewan at Rabbit Lake, Key Lake and Cluff Lake have seen continual advancement of sophisticated exploration models. These deposits result from ancient rapid weathering of a nearby source area which is low in U_3O_8 but becomes concentrated in fast-flowing streams with large boulders downhill from the source area. Ancient or 'paleo' placer deposits are in sandstone sediment in layers of 100 to 1000 metres. Images obtained by the use of EM geophysical instrumentation near the surface aid the hunt for paleoplacer channels. Scintillometers will detect radioactivity. They consist of a crystal scintillator, a photoelectric cell sensitive to the scintillations, and an amplifier. Modelling will produce maps of occurrences and depictions of the areas of interest. A new discovery at Deer Lake in Newfoundland by Spruce Ridge Resources is an example of a promising unconformity uranium deposit.

The Doyon Mine, Quebec.
Cambior

Geological modelling is useful in locating SEDEX or sedimentary exhalative deposits of iron formations laid down in bands. These are formed by the release of ore-bearing hydrothermal fluids into cool bodies of water, usually the sea. Geologists gauge the mineral potential by understanding the spatial correlation between known occurrences and different geological features present in a sedimentary basin. This type of input is well-suited to modelling based on specific computer programs. Another modelling technique involves analysis of brines acquired by drilling to recover the solutions at depth.

Layered intrusions are very attractive for geologists in the search for copper, nickel and platinum group metals. Layered igneous intrusions occur at all levels. Depths at which they were formed depend on the density of the melt, interfaces with the crust, and temperature and viscosity. One of the largest is the Muskox Intrusion in central Nunavut. These intrusions were formed by accumulations of crystals in large magma chambers and appear in funnel-shaped structures. Generally they have sheet-like or 'stratiform' shapes and are host to major economic mineral deposits. They may be displayed by 3-D imaging using the lithology, the nature and composition of rocks that host the intrusions. Traditional exploration tools in the Sudbury nickel camp, such as magnetics and traditional EM geophysical methods, have been successful in locating such intrusions and provide data for their geological modelling. A potential opportunity for exploring layered intrusions lies in hydrocarbon mapping.

Porphyry deposits are the source of 50% to 60% of the world's copper and major sources of other metals, especially gold, so multi-phase porphyry intrusions are of great interest to mineral hunters. Porphyries are relatively young deposits in geological time. They are often circular, sometimes flat, but also tend to slope in a downhill flow. Found in subvolcanic environments, they are radically altered by multiple events, and their chemical and mineral compositions are much changed over time. Such deposits are often massive and amenable to open pit mining giving up high tonnage with low to moderate values per ton. Prospectors may see oxidized, weathered gossan outcrops that are quite common across Canada in nonglaciated areas. These metals are highly magnetic and highly conductive. They may be defined by modern induced polarization (IP). IP and other electromagnetic methods map the variation of electrical conductivity in the earth at low frequencies. When the electrical current is shut off, the voltage decays slowly. The rate of the stored electrical energy discharge is measured by the induced polarization receiver. Both induced polarization and resistivity measurements are taken simultaneously. The numerical data can be used to model 3-D images. Flin Flon, Manitoba is home to Hudson Bay Minerals' flagship operation, Canada's third largest deposit of copper. The deposits there are massive, having been mined since 1927, and there are still more than 21 million tons of reserves remaining.

It would be a gross simplification of the exploration challenge to say that, in Canada, diamonds are hosted by igneous rocks. Kimberlite

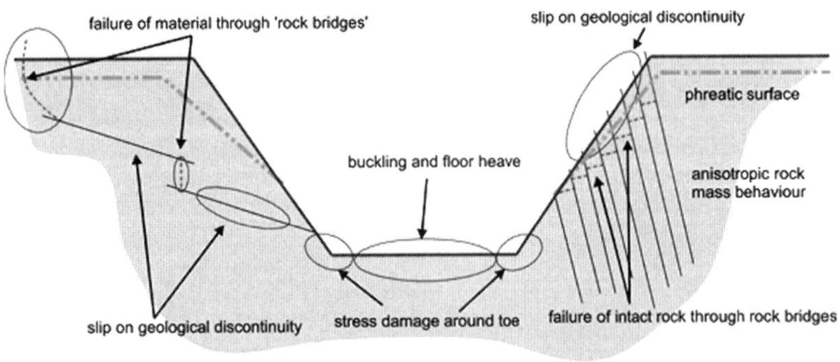

Open pit modelling
Beck Arndt

pipes and sills bring them to surface from depths greater than 150 kilometres. The odds against locating the precious stones are significant because economic concentrations of diamonds are thought to be in only 1% of the kimberlites worldwide. The crystals are incorporated in magma during its ascent to near surface. Diamond hunters find their quarry in depressions around lakes or swamps and rarely in outcrops, due to erosion. The most prominent indicator minerals are garnets. Data for geophysical mapping are gathered by aeromagnetic surveys and aerophoto surveys. These processes are expensive and include LANDSAT satellite remote sensing: geophysical surveys of a wide variety including magnetic, gravity, electrical, radiometric and seismic profiles.

In the central James Bay Lowlands, aeromagnetic surveys with fixed wing aircraft flying at altitudes as low as 30 metres, enabled by the limited tree cover, have been successful. At such altitudes the magnetic amplitude is significant and shorter wavelength characteristics are evident. This has been followed up by detailed magnetic mapping using helicopters. Another hot area of Canada for diamond exploration has been in the Buffalo Head Hills of Alberta. There are kimberlite outcrops near Red Earth Creek. At least twenty-four diamondiferous kimberlite pipes have been located there. The Geological Survey of Alberta has mapped distribution patterns and made summary maps as well as structural-emplacement models.

Dikes in the Arctic require much planning and care in construction.
Diavik

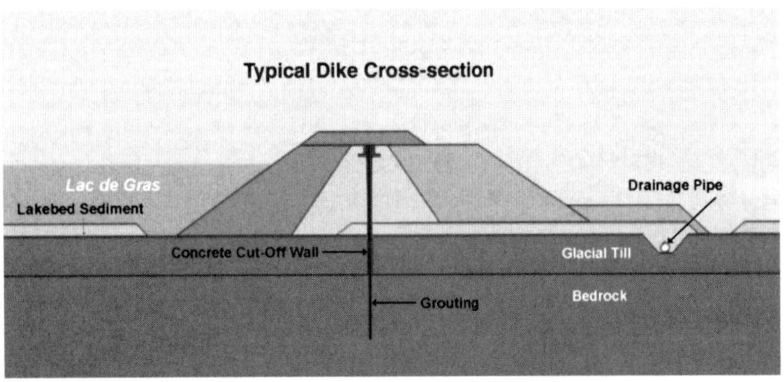

The PDAC has pushed for cooperative geological mapping strategies in Canada to the tune of an annual expenditure of $50 million per year over ten years. Only governments at all levels have the heft to enable the rebuilding of Canada's mapping infrastructure.

Another modelling tool in the detection and identification of minerals is hyperspectral remote sensing or imaging spectroscopy. This relatively new technology includes large data sets with the spatial information collected as a data cube. The imagery can be visualized, processed and analyzed using patented image processing systems.

First shipment of concentrate from Voisey's Bay

Newfoundland Natural Resources

The impact of NI 43-101 on reporting ore reserves has been described elsewhere. The system was set up due to the Bre-X scandal which destroyed mine financing potential for ten years. Geological tonnage and grade calculations are one of many initial indications of ore reserves which must be certified by a 'qualified person.' This chapter might well conclude by examining the certificate provided recently by an exploration geologist acting as a qualified person. The data contains his qualifications, years of professional experience, professional memberships, familiarity with the area of interest, the sources of data for his exploration report and a description of any options he owns in the company. Such a certificate should inspire public confidence, as qualified person can point to experience, a track record written in metal. In the final analysis, only the amount of metal collected at the end of the milling process gives an accurate measure of the ore reserves outlined.

Applied Science Goes Prospecting

Geophysics

A good ore body is like a beautiful woman. It reveals its secrets slowly.

Terry Crebs, Voisey's Bay geophysicist in *The Big Score*

They used to call geophysics the 'black box,' for there was more than a hint of sorcery about this application of science to prospecting. It had its genesis long before World War II, and some old-timers were not convinced that scientific methods were the answer to all prospecting requirements. Manitouwadge camp discoverer Roy Barker averred to all and sundry at an industry event that he had his own black box. Eager inquirers watched as he thrust his hand into the box. "This is an instrument that has helped in the establishment of practically every mine in Canada and it can be used to find many more," said the veteran prospector. Like a magician pulling a rabbit out of a hat, he drew out a well-worn prospecting pick and brandished it above his head. "And the way to use it is to crack rock!"

Early days, operating a "black box"
Irwin Hoffman

Just a few years after Barker amused people at the PDAC convention in 1954 with his hammer and box show, Texas Gulf Sulphur made a world class base metals discovery near Timmins, Ontario, and the familiar pick was not used in exploration at the overburden-covered site. Instead the find came after six years of advance planning, geological modelling and expensive airborne electromagnetic system surveys using helicopters.

Geophysics uses measurements of physical properties combined with geological information to infer the earth's internal composition. Heat flow, geomagnetic and gravity fields, electrical conductivity and information obtained from seismic data are all grist to the geophysicist's mill. The science has developed over many years. Edison used a magnetometer in Sudbury in 1906 in a search for metals. Hans Lundberg of Sweden made the first aerial geophysical readings and located two base metals deposits in the Buchans area of Newfoundland in 1926.

The use of geophysical methods gained a huge boost in research during World War II, when magnetic detectors were used to locate enemy guns, submarines and aircraft.

Geiger counters and scintillometers were used for hunting uranium in the mid 1950s.
OMNDM

Perhaps the most noted Canadian pioneer in geophysics is Dr. Norman Keevil who held the first chair in geophysics at the University of Toronto in 1937. He went into business for himself after the war, and using the newly-developed airborne magnetic detector he discovered the largest body of almost pure chalcopyrite ever found in Canada and the rich copper mine that came out of it at Temagami, Ontario. He also financed other ventures including what became the Teck Corporation.

In addition to Keevil's work, private airborne research discovered the iron ore body that became the Adams Mine near Kirkland Lake. Inco used flights with a magnetometer to check its own knowledge of local deposits in the Sudbury basin. Over the next few years such instruments as the gravimeter, the Geiger counter, the scintillometer, the gamma ray spectrometer and various electromagnetic survey systems were developed and later refined. McPhar Geophysics and Sharp Instruments are companies, launched by Canadian war veterans and entrepreneurial geophysicists after 1946, which are still in operation. There are many present-day airborne and ground geophysical contractors worldwide.

Today the AEM or the airborne magnetic survey is a common exploration tool. A recent advertisement placed in *The Northern Miner* by the government of the Northwest Territories sought companies interested in doing "a digitally-recorded, high sensitive magnetic survey in Nonacho Basin and to compile and package the acquired data into digital line and grid archives and gridded, contoured maps." There had to be several years of AEM successes before government readily acknowledged the benefits of such surveys. Examples of a few early discoveries based on AEM data were the Heath Steele deposits in New Brunswick; the Mattagami ore body at Matagami, Quebec; the magnetic ore body at Marmora, Ontario; the Pyramid ore bodies of Pine Point, Northwest Territories (N.W.T.); the Sturgeon Lake properties of Mattagami Mines; the Uchi Lake ore body in Ontario; and the occurrence at Ruttan Lake owned by Sherritt-Gordon Mines in Manitoba.

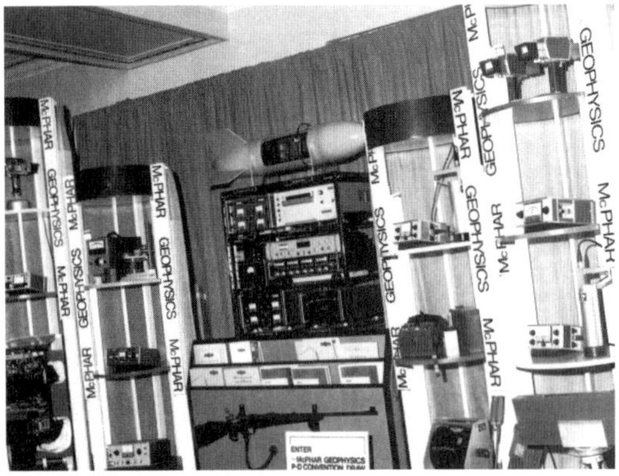

McPhar used a give-away rifle as a crowd stopper at the 1973 trade show.
McPhar

The commercialization of airborne geophysical surveys has come a long way since the days when a primary consideration had to be to ensure a balance between the desired geophysical data results and safe and effective aircraft and helicopter operation. Once it was found that mapping could be done so much faster by aerial means than with regular field methods, interest among geologists and prospectors in the science grew. The aeromagnetic anomaly near Marmora, Ontario, was charted in 1950. The ground had previously been masked by a cover of sediments. Within five years of this mapping, which showed a striking view of the anomaly among other grid lines in the surrounding country, a huge pit was developed that produced seven million tons of iron pellets for Bethlehem Steel in its first fifteen years of operation.

An 'ABEM WAD1' in use in the bush
Atlas Copco

The first ground EM system called the vertial loop was developed by Inco after World War II under the direction of their geophysicist, Dr. Stanley Davidson. McPhar Geosurveys Ltd. claims the honour of designing and building the first commercial ground

EM system way back in 1950. Today the company brochure offers interpreting and modelling of data in a series of maps which help delineate geologic and economic targets for its clients. Geotech is a company which has a good history of development of advanced digital helicopter-borne electromagnetic systems. This firm relies on its 'versatile time-domain' EM system which enables penetration to 800 metres below surface. Veteran Fugro Airborne Surveys has profited from the advances in airborne geophysics, but its people admit the cyclical nature of the business. In the mid-eighties cutbacks in exploration budgets resulted in a real downturn for this scientific aid to prospecting. Fortunately the first decade of the twenty-first century has seen a real upswing in AEM exploration activity.

Mine finders Don McKinnon, Fenton Scott and Peter Ferderber use a 'doodlebug' in the late sixties.

McKinnon Prospecting

In all geophysical surveys, the target is an anomaly, a location where the earth exhibits irregular background physical properties. Over the past six decades there has been a rapid development of ground surveys with a wide variety of techniques to choose from. All are designed to provide a subsurface 'picture' of the earth.

Magnetic surveys gauge the strength of the earth's magnetic field and variations may be measured by a magnetometer. Magnetic minerals associated with iron and nickel show up as an anomaly. The data provided in such surveys may be used in geological mapping. Contour maps and other graphic profiles are generated as required. These surveys are rapid and generally efficient.

Electromagnetic methods have been useful but had a bulky start. As early as 1946 a device was developed which consisted of two instrument cabooses pulled by tractors. An operator pointed a vertical transmitter coil at a receiver on the other caboose. The thing was called a catamite, or 'devil's minion.' The name was appropriate if only for its awkward means of operation. EM only became useful when the process could be substantially downsized.

The EM principle was soon bulked down. The method detected all conductive bodies, not just important mineralization. Feeding an alternating

A geophysicist works in the west Raglan area, Labrador.
Knight Resources

current into a wire coil aligned in the required direction will produce an alternating magnetic field that induces a current in any adjacent electrical conductors. The resulting magnetic fields are measured by a search coil attached to a voltage meter. The trouble is that the conductor could be a handsome deposit of metal sulphides, or perhaps a conductive field due to graphite or barren sulphides. This method originally gave a picture of perhaps 100 metres below surface, but since then low frequency 'pulse' EM methods have greatly increased the depth penetration.

Another aid in exploration is electrical resistivity or conductivity. An artificial source of current is introduced into the ground through a pair of electrodes. The current flows between these electrodes and fans out in a pattern; the intensity depends on the conductivity of the ground. The operator uses a second pair of electrodes to measure the voltage pattern along a traverse or straight line and keeps moving the electrodes. An ore body containing metallic minerals will show up as anomalously low resistance.

Unmanned aerial vehicles collect data for the mining industry. Pictured are Declan Sweeny, Buddy Doyle, Alan Wood and Andrew Kondor.
Universal Wing

Induced polarization or IP is a method developed in the late fifties and is now one of the most commonly used forms of geophysical exploration in the world. The idea is similar to resistivity in data collection but involves analysis of the length of time the earth remains disturbed electrically after the disturbing function has been removed. There is a brief storage of energy that can be measured. This method is most useful in sulphide exploration. Generally both IP and resistivity measurements are taken at the same time. The measurements are not taken over frozen ground as good contact with the surface is essential.

Gravity prospecting involves the measurement of variations in the gravitational field of the earth. This means of obtaining data depends on the fact that the force of gravity is stronger in dense rocks than in those less dense. Even minute changes may be measured by extremely sensitive, and expensive, instruments. This form of prospecting is especially useful for detailed follow-up of magnetic and electro-magnetic anomalies during integrated base metal surveys. Ore bodies are very dense and are most pronounced in gravity surveys.

Airbourne survey to identify kimberlite targets

Sask. Industry

Downhole geophysics is welcomed by drilling firms and geophysical companies as another service they can offer in the exploration field. Many technologies using downhole measurements obtain higher data resolution. The effects of near-surface signal attenuation and some forms of noise are decreased. The driller's role is to see that the hole is stable and the casing is in firm contact with the rock mass. Electrical means such as resistivity work well and conductive bodies may be recognized with downhole exploration.

Research in geophysics has generally been funded through universities but private contractors are deeply involved in research and development. Scintrex is possibly the largest company designing and manufacturing geophysical surveying instruments. Such firms spend a large percentage of sales revenue on research and development. They employ highly educated scientists at the master's and doctoral levels. High speed multicomputer surveys have proved most useful in seismic exploration methods. The idea stems from earthquake seismology and can locate features such as buried channels where heavy minerals may be accumulated. In earthquakes the earth is fractured and rocks on opposite sides of the fracture move in relation to each other. Anomalies may be observed due to the reflective pattern from potential ore deposits and their enclosing rocks. The waves must be monitored continuously, hence the utility of computers receiving the data from sensitive geophone receivers located along the survey line.

Exploring for uranium in Saskatchewan

Forum Uranium

The Titan 24 technology developed by Quantec Geoscience offers deep electrical earth imaging which can identify conductive areas from surface to 1.5 kilometre depths and beyond. Structural information and conductive bodies are modeled in 3-D, thus greatly increasing the chances for drilling success. In historic mining camps the process may also be used to meld historic and new data to extend and predict new reserves and resources. Companies that thoroughly image their ground area can save large sums in drilling costs. Companies that have pioneered the use of Titan 24 technology are Goldcorp, FNX Mining and Xstrata. One junior, Tribute Minerals, has made extensive use of this 3-D modelling mainly due to beneficial investment by senior companies such as Barrick and Noranda.

Magnetic surveying in the Northwest Territories

Southern Era

Three dimensional visualization and modelling and deeper penetrating geophysical information have been a boon to efficient mining. The latest geophysical subsurface imaging capabilities are so well-developed that in 2004 the Chairman of giant miner Freeport-McMoran could state that the next world class ore body will not be found by digging a trench. Valuable exploration dollars spent on obtaining information at depth reduces timelines for project execution. In the future 'elephant' ore bodies will be located deep in the earth.

Applied science has been shown over the past six decades to be not the black box that appeared to threaten prospectors but one of the best tools in the exploration arsenal. The geophysicist of today has advanced degrees and is required to consult with project geologists, working at planning, supervising, processing and interpreting surveys. This modern mining explorer is a certified professional geoscientist and the designation 'P.Geol' proclaims one who is devoted to continuous learning and professional development. The hammer alone does not cut it anymore.

Analyzing the Earth
Geochemistry

The only way of discovering the limits of the possible is to venture a little way past them into the impossible.

Arthur C. Clarke

A prospector working through the bush one time may have come across a stand of poplars with yellow leaves instead of green. This might have prompted him to try to discover what existed in the surrounding soil to produce this anomaly. Unwittingly, he was practicing a science that has grown up alongside geophysics. This is the discipline of geochemistry, the science that studies the distribution of chemical elements in nature.

Vegetation, soils, rocks, silt in streams and even running water all come under the purview of the geochemist. Old timers used signs, colours and taste. Small Arctic carnations were sometimes called the copper flower because variations in their colour could indicate the presence of copper. Spectra-analysis of heather from the moors of Cornwall has been used to locate tin. The modern geoscientist uses advanced variations of the same means. Analytical techniques used in this field sometimes lead to mineral deposits. But geochemical analysis is an indicator. The exact position of minerals, their dimensions and depth will still have to be outlined by conventional means, likely diamond drilling.

Except for some high-tech equipment, the process of fire assaying for gold has not changed since long before the end of World War II. Nitric acid is used to dissolve any silver present in the sample. The gold is then dissolved by adding a concentrated solution of hydrochloric acid to form aqua regia or 'royal water.' The most precise instrument available today to measure a test tube of this solution is an AAS or atomic absorption

Stream sampling with helicopter assistance, Cochrane, Ontario
Expedition Helicopters

spectrometer. This computer-controlled device is programmed to measure gold standards. A flame fed by an air-acetylene blend is ignited in the chamber next to the computer. A tiny tube sucks up the gold in the solution and the flame changes as the liquid is vaporized. Set in the lamp housing of the AAS is a gold lamp which has a coating of the precious metal within its core. The light from the lamp interacts with the gold atoms in the flame and the instrument reads the results. The gold in solution can be detected in amounts as small as 0.0005 ounces to the ton.

Stream sampling in Northern Ontario
Southern Era

Traditional fire assaying gained a boost with the advent of the atomic absorption spectrometer. There are several high tech alternatives for analysis out on the market now. ICP multi-element or Inductively Coupled Plasma is a method for detecting trace metals in environmental samples. Mining and exploration companies are very sensitive to public opinion in dealing with the environment, and advances in metal detection technology have reduced environmental impacts. The goal of ICP is to get elements to emit characteristic wavelength-specific light which can then be measured. Plasma is a gas in which atoms are present in an ionized state. An ion is simply an atom or group of atoms which has lost or gained one or more electrons. This analytical tool generates plasma. Argon gas is used to convey the sample to the plasma.

Another acronym well-known in geochemistry is NAA or Neutron Activation Analysis. This is a technique which very accurately determines the concentration of elements in a sample and has the advantage that it does not destroy the sample in the process. This is isotope geochemistry. The radiation field of a nuclear reactor is used to bombard the sample with neutrons. The elements thus 'zapped' form radioactive isotopes. Since the radioactive emissions and decay paths for each element are well-known, this data enables a study of the spectra of emissions in the sample and determines the concentration of the elements within it.

A mining process which benefits from the work of geochemists is cyanidation, by which crushed gold ore is treated with an aqueous cyanide solution in air to dissolve the precious metal. Pure gold is recovered by reduction of the ion to the metal. Traditional methods of gold recovery have been relatively inefficient. In addition, there are environmental risks connected with cyanide, so much so that the United Nations Environment Programme has developed an international code for the management of cyanide. Geochemists and metallurgists must work together to devise a plan for every mining operation. This will include examination of the chemistry of the cyanide, its toxicity, and mine water and wastewater management. The InLine Leach Reactor is a popular method of producing gold concentrates as it uses a simple leach process to recover over 95% of the gold.

One new advanced surface exploration tool in the geochemist's arsenal is MMI or Mobile Metal Ion soil geochemistry. This may be used for gold and other metals of interest. Measurements in this exercise are taken 10-30 cms below surface. A selected line is designed to traverse the location of a relatively small seismic explosion. Loosely bound elements are detached from the soil after having traveled upward to the surface. In the laboratory they undergo a chemical extraction process that strips out only the most weakly bonded metals. Sensitive instruments analyze the small concentration of mobile metals, and their location may be mapped where promising mineralization is indicated.

Till sampling for diamonds
Southern Era

We are accustomed to the expression that people are "hitting the trail," a reference to pioneer days now entrenched in the language. The exploration geochemist is also interested in trails, only these routes go back many thousands of years. There is an important distinction between primary and secondary chemical distribution of elements. The former is centred on the location where the ore was originally formed while in the latter the ore of interest winds up some distance away from its birthplace. Movement of ground or water will sometimes move elements far from their original location where they were laid down. The geochemist looks for clues to locate the path the elements

Mapping a trench at Cape Ray
Cornerstone Resources

followed. Samples are carefully mapped. In the case of glacial action, the observer notes how materials have been deposited in the wake of advancing or receding glaciers. Traces of diamonds may be discovered by means of such exploration. Secondary distributions may occur in other minerals such as copper. Copper is heavy but the huge mass of the glaciers had no problem transporting minerals or anything else in their path. The geochemist must have knowledge of geological processes that cause such dispersions in his exploration arsenal.

Pressure washing an outcrop at Cape Ray
Cornerstone Resources

In common with most areas of exploration activity, mineral sampling is expensive and must be uniform; otherwise the geochemist is in the unfortunate position of comparing apples to oranges. Whenever clients find their series of analyses provide unexpected results, the culprit is almost always poor sampling methods. Orientation surveys are essential to a geochemical exploration program. Each site must be evaluated in the context of the information sought. The careful searcher has a plan which includes the sample density, preparation of samples in the field and the laboratory, data recorded at the site and the assay methods required. Rock chips are more representative of the average grade than grab samples of individual pieces of rock. The investigator will weigh the options in several sampling mediums. In nature's own sampling, mineral signatures are often very reliable in glacial till or moraine.

Geochemists are on the scene at the beginning and at the eventual end of the mining story. This is a most important factor in the mine operator's protection of the environment. The specialist assesses, predicts and controls water-borne migration or mine site drainage. Exposure of primary minerals to air and water can result in leaching of metals. In the presence of sulphide minerals, an unfortunate result could be the generation of acidity. All areas of operations merit geochemical attention. Pits, underground workings, waste dumps, tailings and even broken-rock building foundations are examined for possible deleterious environmental effects. No harmful discharge is acceptable and it is the job of the geochemist to see that such events do not occur.

Fisheries biologist Genevieve Morinville at EKATI Diamond Mine

Rider Petch

The trend in geochemistry has been towards the standardization of laboratories and accepted procedures for analysis. As in other areas of the industry, analytical firms are consolidating and establishing regional centres. Apart from the need to provide timely service to client companies, one main driving force in this movement is the sheer cost of capital investment in analytical equipment. The expense of research has to be largely financed by government through the universities, although large private companies are contributing substantially.

The mountains are so vast and my exploration plane is so small!

Sander geophysics

Geochemistry needs no testimonials because the science is now considered vital to exploration and analytical success. Kennecott was one of the first companies to find new sources of metals by scientific means, locating the oxidized, weathered and decomposed rock of a gossan by geochemical tracking in 1953. In the sixties and seventies geochemists located gold and silver in B.C. using methods that were really an extension of stream panning and soil sampling of past years. In 1963 a group led by Karl Springer found geochemical anomalies in Newfoundland near Daniels Harbour which led to a zinc mine a few years later. Today geochemists routinely reconnoiter stream sediment and soil because porphyry deposits have large chemical haloes. Hydrothermal fluids offer clues to VMS deposits.

Geochemistry has advanced since World War II to the point that doctorates are now offered in this vital analytical science. The perspective of the geoscientist is that the present is the key to the past and the past is the key to the future.

Pictures from on High

Satellite Imagery

If we learn one thing from the history of invention and discovery, it is that in the long run — and often in the short one — the most daring prophecies seem laughably conservative.

Arthur C. Clarke, 1951

The first remote sensing experiment was accomplished in the late nineteenth century with the aid of cameras tied to balloons. Photography was used to identify enemy movements and locations during both World Wars. Only after the launching of the football-sized Russian Sputnik satellite in the late fifties did the major powers begin to realize the potential of remote sensing. Satellites have a romance to them in public opinion but become somewhat demystified when described as specialized wireless transmitters placed in orbit round the earth. The first satellite pictures of the earth were made by the weather satellite TIROS-1 in 1960. The pictures were grainy and of poor quality, but as the technology developed and the orbiters grew in size, image quality was progressively enhanced until today even citizens sitting at home with their computers can access Google satellite image maps offering highly detailed and precise images.

Atlas Copco signifies its world reach.
Atlas Copco

While industry is sometimes slow to react to the opportunities offered by emerging technology, in the case of images accessed from space, the federal government has been quick to respond. The Americans offered to share the product of their ERTS (Earth Resources Technology

Satellite) in 1972 and Canada was the first nation to set up its own facilities to receive and process the data from the U.S. venture. People in all fields of endeavour were quick to realize the utility of remote sensing. Enterprising miners noted that geological structures could be easily examined using images from above. Both Canada and the United States began to cooperate almost immediately on a satellite triangulation survey. The geodetic survey became the starting point for all other surveys. Mapping and the study of terrain were just a couple of the advantages the eye in the sky could offer the nation. The development of the high Arctic could be understood and managed making use of remote sensing.

The satellites which grew out of early weather instruments are marvels of technology but it is the tools they offer clients which have fired the development of remote sensing. Simple photography is just one example. The birds detect and document electromagnetic radiation that is emitted or reflected from the earth's surface. Appropriately designed spectral sensors can detect wavelength characteristics of electromagnetic radiation. Visible light is recorded on black and white photographic film. Light from the infrared and thermal parts of the spectrum is recorded on colour infrared film and thermal scanners. Multispectral scanners (MSS) then detect energy electrically and convert the signal to digital form. These forms of remote sensing were the forerunners of satellite means of reporting to ground stations.

The tiny football-size Sputnik started the space race.

Two early pioneers in the field of remote sensing are RADARSAT and LANDSAT1. The former is an earth-imaging Synthetic Aperture Radar (SAR) satellite system. This is not affected by weather, atmospheric or light conditions. RADARSAT has seven SAR imaging options and is useful in the mining industry for observation of land use. The LANDSAT1 program followed on from the ERTS system mentioned earlier. The Apollo moon missions inspired the LANDSAT idea to photograph the earth's land surface from space. The purpose of the satellite was to inventory and monitor the planet's landmasses. At first the camera developed problems in altitude control but the MSS performed flawlessly. LANSAT1 had acquired over 300,000 images of the earth's land surfaces before it ceased to operate in 1978. Since that time, there have been

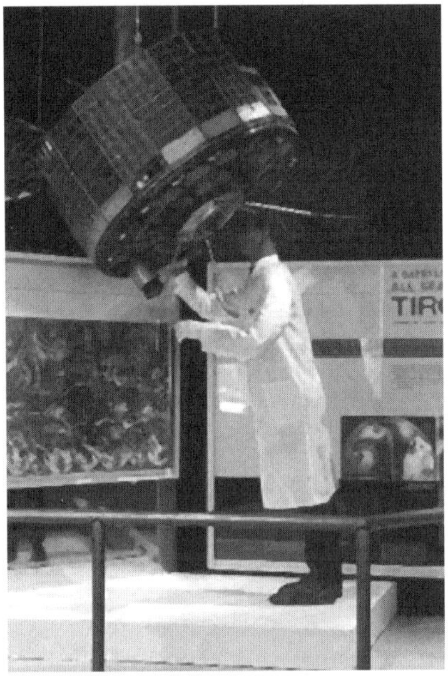

A technician checks the TYROS satellite.
Smithsonian Institute

RADARSAT on display
in Ottawa

six more satellites in the program. Over three decades of research have produced satellite mounted-cameras capable of delivering images of extremely high intensity.

Early fears of speeders being monitored on highways by birds overhead have proved baseless. However, photographs taken in space are now able to show stationary cars clearly. The cameras that one scientist described as "paparazzi in the sky" have not invaded privacy but rather have enhanced our understanding of the earth. The most long-lived government-funded satellite program remains LANDSAT, now working with its seventh bird. LANDSAT7, launched in 1999, offers six-band multispectral imagery: three in the infrared spectrum, two in the thermal spectrum and one panchromatic band. The vertically-collected imagery captures the same area every sixteen days. The Geoimage Company processes the LANDSAT imagery to predict mineral deposits, and its hardcopy prints assist in mineral exploration and planning. Although still a costly exploration expense, the price of satellite data has dropped considerably in recent years. The big bird gives value for money. LANDSAT7 has been called the best, most stable earth observation instrument ever placed in orbit. The United States has ensured the project will continue with its LDCM or Landsat Data Continuity Mission.

Photographs and other images from both aircraft and satellites can be most useful in exploration and evaluating the resource. Infrared imagery will reveal geological faults, and outcrops and vegetation overlay can give clues to mineralization. Radar imaging is an aid to evaluating structural geology. LiDAR is a source which uses lasers rather than radar waves, enabling geologists to 'see' through forest cover and get at the bare earth topography. Much satellite imagery is archived and thus cost-effective in evaluating potential properties. Such libraries of images assist in field mapping and guiding field geologists. Planning of potential environmental measures and remediation are another way in which the eye in the sky may be of service to mining companies.

Custom imagery is more expensive than archived material, but with the number of commercial satellites now in operation, the data market is now highly competitive, allowing clients to shop around. Job specific contracts ensure that imagery that meets particular needs for resolution and time of year. Satellites have a particular advantage over aircraft during the exploration phase. Pilots and survey parties may leak information, but the satellite service provides anonymity. The required data is usually processed using a mining consulting firm with expertise in remote sensing.

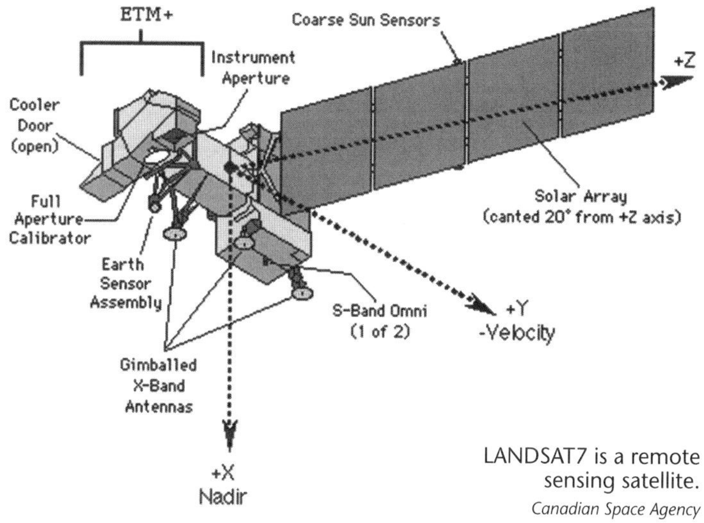

LANDSAT7 is a remote sensing satellite.
Canadian Space Agency

Satellite exploration has been heavily funded in recent years by central governments. Chief among these are Australia, the United States and Russia. Other countries such as France and Britain are also involved in obtaining data from space, and smaller nations can simply buy satellite time and expertise. Natural Resources Canada has a newly funded portal called GeoBase which will provide a closer look at the country at a higher resolution than enjoyed previously. The Canada Centre for Remote Sensing is an arm of the Earth Sciences Sector of Natural Resources Canada. Simply put, this agency's job is to receive data, process, archive and disseminate the material. The CCRS deals with geomatics, the science of using mathematical methods on data about the earth's surface.

A complete topographical coverage of northern Canada is essential for establishing a base for resource development, an initiative, obviously, of great interest to the mining industry. In 2003 it was estimated that the unmapped areas of the Canadian North would cover 1500 topographic sheets at the 1: 50 000 scale. Data required includes planimetry, which describes the earth's features, and altimetry, which portrays the earth's relief. RADARSAT and LANDSAT7 have been engaged by Natural Resources Canada to produce data sets, as well as conventional stereo digitizing for data validation purposes. Some of the data received has been like the curate's egg, good in parts, so there has had to be ongoing fine-tuning of the project specifications. The work along these lines being done in Nunavut is called the CartoNord Project.

There has been a rapid growth in the last few years of private commercial data banks. The original national players in the space information and observation business now have to share the field with what one writer has called "a fleet of mercantile sentinels" which provides high-resolution imagery to customers around the world. Russian intercontinental ballistic missiles are in the satellite delivery business. One client of this private enterprise space exploration has been Earth Watch Inc., an American firm. But risks are enormous in this business, and its first Early Bird 1 satellite failed to operate satisfactorily. One firm, Space Imaging, contracted with Lockheed Martin to build its satellite and the IKONOS satellite was provided by Eastman Kodak. A whole range of resellers has sprung up — companies which contract for satellite data and then interpret and provide the product to client companies, including the larger world mining companies. A firm mentioned earlier providing both data bank and on-demand images is Geoimage, an Australian venture which uses archived material from the SPOT Company. Geoimage acts as a business partner with other commercial satellite companies. One executive with one of the big commercial remote sensing operations has been asked if government has better imaging systems than the private sector. His answer was that he did not know and that even if he did he could not share that information.

To bring us down to earth, no ore deposit has been found *directly* by remote sensing. Put together laboratory work and ground observation, add basic geological concepts and combine with satellite-collected data, and you have a good chance of locating promising resources. Typically, on the Canadian Shield, ore engaged with the assistance of satellite data will have good geophysical response characteristics. On the other hand, in exploration for uranium ores, promising data may be the result of granite and metamorphic wall rocks that are radioactive, resulting in a hash of background signals and false anomalies that airborne surveys fail to delineate. Best electromagnetic responses are gained from massive sulphides and yet variable amounts of pyrite and graphite often give a geophysical response that cannot be distinguished from the ore itself.

The federal government is convinced of the variety of benefits for many disciplines of the advanced technologies achieved by remote sensing. Unfortunately many mining companies have not taken advantage of this major resource which to their potential benefit is so hugely subsidized by central and provincial governments. One problem may be that clients are unfamiliar with the science and see the cost as prohibitively high. Australia's Immersive Technologies has countered the former difficulty by producing computer-based training for the industry. Currently newer high-resolution imagery in selected spectral bands opens up areas such as surface lithologies, geological structures and fracture systems to inspection. Instead of making an initial onsite survey, an accurate, detailed 3-D visual model of the site can be constructed with information obtained by the high resolution imagery when linked to digital terrain models (DTM) and specific elevations.

As mining people become more comfortable and accustomed to the tools offered by satellite observation for exploration and other phases of the mining sequence, promising locales may be explored in simulation exercises, which much reduce the investment risks associated with real-world mining.

Getting at Fortunes in the Ground

Equipment at Hand

His rod consisted of two whalebone strips,
Bound together at the outer tips.
With this device he could uncover veins,
Thus obviating all prospecting pains.
When, half-deceiving and half deceived,
He passed away, I doubt if any grieved.
His day had passed, his type lives forever;
Though I don't think the present kind so clever.

Canadian Mining Journal, March 1923

What follows should properly be called Auxiliary Supportive Equipment, but the title does not sing, though it refers to ways of locating minerals in place, extracting this bounty of the earth and processing what has been recovered. The ore body in question may not really qualify as a bonanza — but then miners are always optimists and hope can fuel a lot of action.

There are three major processes that make use of equipment: these are diamond drilling, which is generally agreed to account for half of most exploration budgets; the extraction of the minerals; and the judicious application of metallurgy or, as it is sometimes known, minerals engineering.

Clayton Parsons is on his way to a bid meeting called by one of the two major nickel miners in the Sudbury basin. As area manager for Major Drilling Group, he covers Sudbury, as well as the Atlantic provinces. The latter responsibility is in part because he is a Newfoundlander and knows the ground. Quite a few drillers come from the Atlantic provinces. Fishermen adapt well to the trade because they are used to a demanding outdoor life of long hours and isolation, and they are handy with tools and machinery. They have a saying about bad weather: "It is a cold day when your teeth are chattering and they are still on the night table." A fisherman who adapted well to drilling, Parsons ascended the employment ladder through helper, operator and foreman before being promoted to his management job.

Major Drilling is a big international company and does one-sixth of all the diamond drilling in the world. The big garage at the Sudbury office has tools and other equipment but all the drills are out on jobs. Despite the competition among drilling companies in Northern Ontario, any firm with decent equipment and skilled workers has all the projects it can handle in the current phase of the mining cycle. Out in the front office a slender pipe, perhaps eighteen inches long, crammed with electronic equipment maintains a prominent place set in a glass case on the wall. This is a Reflex EMS drilling surveying system, a tool used to measure the deviation in direction of a drill hole. At a price tag of around $18,000, it is most handy in the effort to keep drill holes spot on the angle and direction requested by the client. A closer look reveals that the delicate cylinder has been slightly bent. Run over by a truck on site and now useless, it is a constant reminder that drilling is a costly business and proper handling and storage is essential.

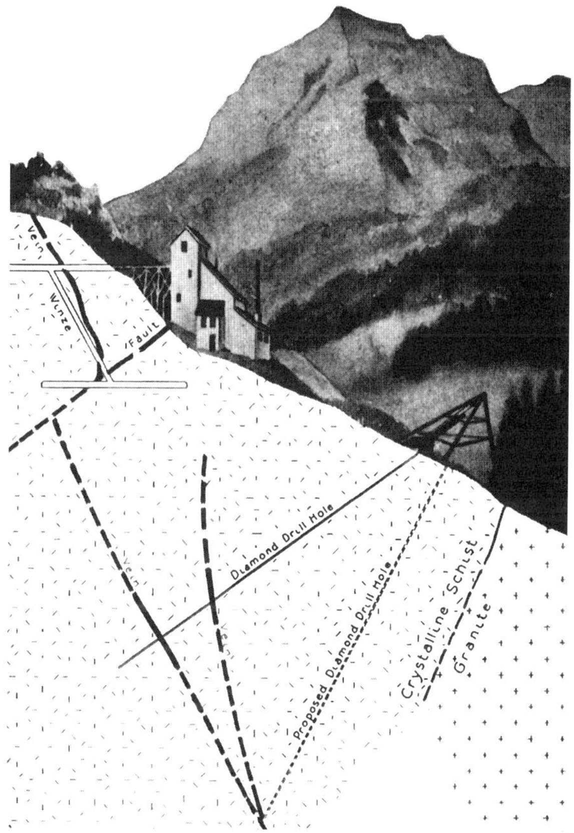

Diamond drilling locates a vein displaced by a fault.
Boart Longyear

The tender meeting is held in the office of the chief exploration geologist. Three drill company representatives are present as the geologist outlines the location of the proposed holes, drill angles and provides maps to indicate the location. The maps are a great improvement over the directional instructions of the past. They are downloaded from the City of Sudbury website and show the general area but a site inspection is necessary. This involves taking a truck ride, donning orange safety vests as a concession to hunting season and wearing hard hats and boots, even though this is basically a walk in the bush. One of the indicated spots has two access routes. The drillers unanimously turn thumbs down on the first. The hillside that must be negotiated is littered with heavy boulders.

On the way back to the office Parsons is running through the aspects to be considered when submitting his tender. The site is not far from his home base, and the only additional equipment that might be required is a tracked vehicle to pull the drill shack up a small hill. The company wants the work completed before Christmas, so he will need two drills and the equipment schedule will have to be checked for availability of men and machines. The job appears straightforward, but he will still have to build in a contingency amount per metre to take in any number of presently unforeseen events that may slow down the job. The other bidding companies will be making the same determinations.

This is an early depiction of a diamond drill rig but the principle remains the same today.

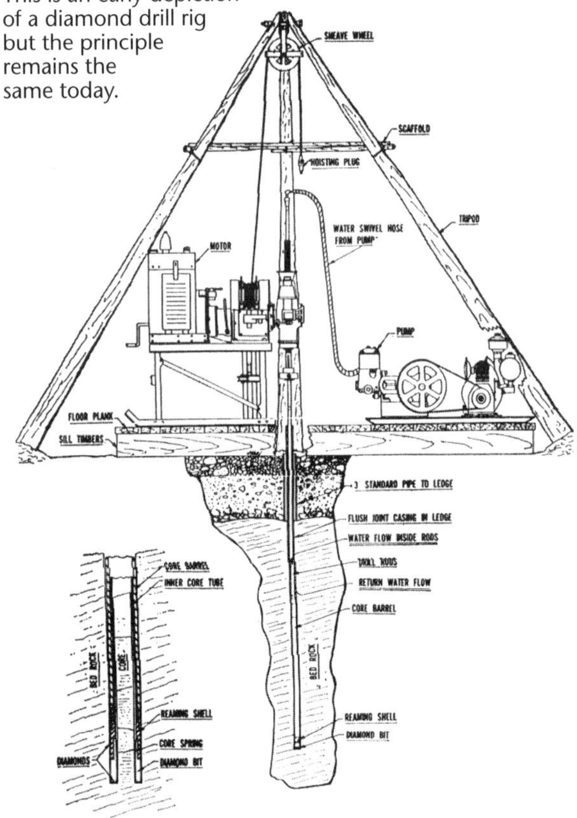

Later that day Parsons will cross downtown Sudbury and pass through Azilda on the way to Onaping. Major has a drill running there, and although the foreman reports daily on the job, the manager likes to visit sites for himself. This one is also relatively easy to access: off a main road and a few hundred yards along a dirt track. Along the way there is a sloop, a long covered steel box on skid runners. This handy piece of equipment is used to transport drill rods to the site. There is a water pipe running down to a nearby lake and a muddy hill for the four-wheel drive truck to navigate.

The drill shack sits in a clearing along with a pile of core boxes, extra rods and other gear. The shack has a lean-to added on, in mining parlance, a 'dry' where the operator and his helper can keep extra clothing and lunches. The men work on twelve hour shifts, 7:00 am to 7:00 pm, so they must prepare for variations in the weather and for food on a long shift. The machinery in the shack gives off a steady roar; the origin of the noise is the main water pump, which puts out 60 gallons a minute at 5000 rpm. The two men welcome the visitor, but any conversation must be carried on outside of the shack as safety regulations require ear defenders against the constant noise.

The shack is open to the sky in the centre to accommodate the drill mechanism, but with the door shut and a small Quebec heater in one corner, fueled by diesel oil, the place is quite warm. As there is always something to do, the operator and his helper have only brief spells of non-activity. There are 26 dials and switches on the control panel and even joy sticks which make operational adjustments much easier. This contract calls for 1400 metres of NQ, which provides for a core of rock 76 mm in diameter (2 3/4 inches for Imperial diehards). The job is down to the 700 metre mark and the core barrel is full; the crew uses an automated rod puller to extract the rock core. An overshot mechanism on the rod is fed down and picks up the head, and at this depth it will take about five minutes to bring the core to surface.

The drill manager notes the good natured bantering between the men and decides all is well on the site. On the way back to the office he estimates how long the remaining work will take. Demounting the drill and its components is faster than the sixteen hours required setting it up. There is another job waiting when the shack, drill and equipment are on the road again.

Famed mine finder Bob Jowsey was once gravely ill and he was asked what would make him feel better. His response was prompt: "Just show me a good drill hole." Diamond drilling has served Canadian mining since 1872 when core was brought up from shallow depths at the Silver Islet mine on Lake Superior. The process has remained functionally the same, and the drill is still the most reliable instrument available to the geologist to physically explore below the surface. The diamond drill secures a core of the earth's crust so that it may be analyzed for the presence of minerals. When a Canadian reference devotes much ink to the "Romance of Diamond Drilling," the practice merits closer examination.

The most comprehensive handbook on the subject of drilling used worldwide is the *Australian Drilling Industry Training Manual*. This heavy

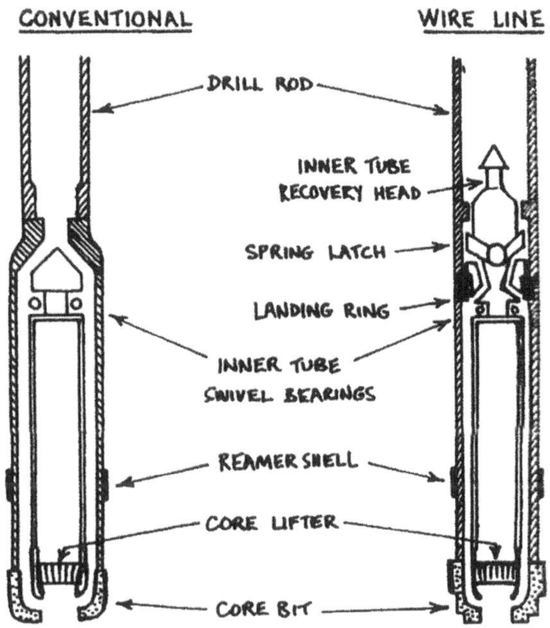

Wireline is the most commonly used drill method used today.
Australian Drill Manual

tome shows how the profile and occupational skills of the driller have developed over the years. Industry recruiters have determined that ideal subjects prefer to work outdoors, enjoy travel and have no problem with living in isolated places. Being on the job for long hours and working under adverse and often challenging conditions do not deter such workers. They must be resourceful tradesmen with a mechanical background and possess the ability to communicate well with a wide variety of persons. A top driller knows all the processes involved in making a hole because each application of the craft is different. The experienced driller is constantly aware of what is happening on the job. The business is one of sensory involvement. He monitors gauges and indicators, feels and listens for sounds and vibration. His knowledge of drilling and circulating fluid keeps the hole open and stable until the job is done. The driller has become a leader and key person in the industry.

One of the great advances in drilling process was introduced in the early sixties. Wireline drilling meant that drill rods no longer had to be pulled out of the hole to access the rock core. Instead, an inner tube containing the core could be pulled to the surface by means of a wire dropped down the string of drill rods. An overshot assembly or latch snaps onto the top of the inner tube containing the core and then the tube is hoisted to surface. Once the core is removed, the inner tube is dropped down the outer core barrel and drilling continues. The wireline process has speeded up drilling and made the driller's work less tiring.

Sixty years ago 30 metres of drilling was considered a good day's work. Now with increases in capacities of drills the output can be 100 to 120 metres of core retrieved in a day. One highly popular rig, commonly called a 'packsack' drill, would never reach such pleasing production depths. This is the Winkie drill now manufactured by Minex International. This lightweight gas powered drill weighs in at only 84.1 kg and may be packed in by two men to remote sites. It has the ability to drill up to 145 metres. As the size of drills has grown over the years, the demand for bigger core samples has led to the introduction of bigger pipe and drill heads. Large drills can recover core of 100 mm or 3.937 inches in width from as deep as 4,500 metres.

An annular diamond-impregnated drill bit does the job of cutting the cylindrical core of rock. For many years the source of the cutting edge was diamonds of less than gem quality. Large scale drilling company Boart Longyear gets its name from stones classified in South Africa as 'bortz.' Then in 1954 General Electric became the first company to successfully manufacture artificial or industrial diamonds. Today they are manufactured in Russia, South Africa and Ireland. Graded as fine to micro-fine, these stones are set within a matrix of varying hardness, ranging from brass to high-grade steel.

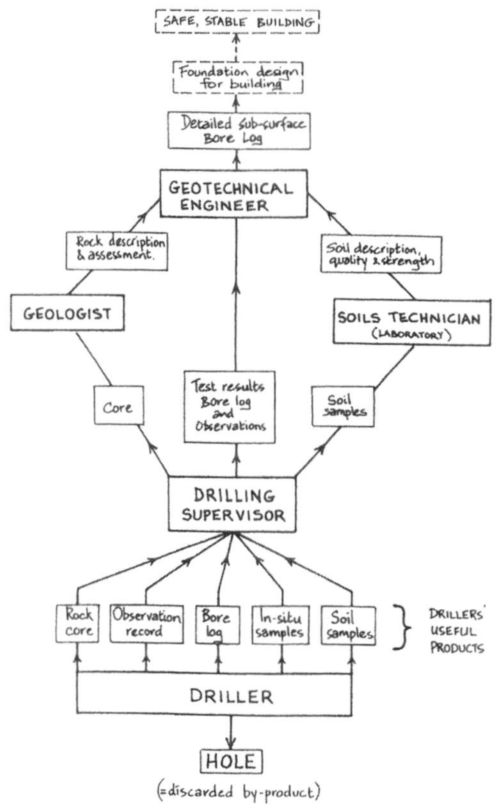

Driller's End Product

Ask a driller like Clayton Parsons and he will say it is not the diamond bits that are the most important items of equipment, it is the rods. Rod condition is one of the many factors of interest to the operator. In deep hole drilling, bad or worn drill rods will result in water lost in the joints, pipe pulling apart in the hole, vibration and even rods stuck in the hole. Nothing can upset drill crews and throw off a drilling schedule more than faulty rods.

The big push in drilling during the past few years has come in environmental protection and worker health and safety. Companies work to reduce their environmental footprint and insist upon a thorough clean up after a job is complete. On all sites the only evidence that a rig had been operating is a capped hole. Innovations such as sonic drilling are done without fluids or mud and there is little disturbance of sensitive sites. So far such methods are only suitable for shallow work. The ***e3*** working group managed by the PDAC, an industry move to establish national guidelines for environmental protection in the mining industry, suggests operating guidelines that may well be the most advanced of any industrial occupation in Canada. Worker health and safety are mandated through extensive training and strong promotion summarized in the telling slogan: "Safety does not hurt." The Canadian Diamond Drilling Association, working with the Prospectors and Developers Association, is a strong voice in safety and health, commissioning studies on a variety of issues, such as noise and vibration reduction As well, the widespread use of mechanized rod handlers has reduced injuries to the hands and reduced operator fatigue.

The highly portable Winkie drill is efficient as well as having a catchy name.
Minex International

The drilling industry has continually moved with the times. The widespread availability of helicopters has resulted in the redesign of drills and rigs so that equipment is produced in modular form for ease of transport. This portability also aids in ease of taking down equipment for more rapid movement to a new site. Companies offer additional services to complement the drilling such as geophysical work. But, for all the moves to establish common core training programs, the biggest problem facing the industry is the availability of labour. The boom and bust nature of mining in Canada for several decades has meant layoffs with the result that skilled workers have left for employment in other industries. The numerous advertisements for skilled drillers every week in *The Northern Miner* attest to the sparse labour pool.

Advances in drilling have been sparked by the oil industry. There are many reasons why drill holes deviate from the original plan. Flexibility of drill rods, reaction to rock structures and the fact that inclined holes can deviate upwards as drill rods lie against the bottom of the bore all contribute to the misalignment of holes. The oil industry led the way in using wedges to control deviation, and this has become a common practice. A new method of directional drilling developed by Norwegian company Devico was first used at the former Dickinson Mine at Red Lake in the late eighties.

This allows for the recovery of a core while deflecting a borehole at a controlled rate. With this system multiple deviations may be made in sequence.

A stationary outer barrel controls the directional setting while the inner core barrel rotates. The outer barrel has a locking piston which is not released until the tool face setting is correct. A packer assembly causes guide bars to penetrate the borehole wall and prevents the outer casing from rotating. Finally what is known as an 'eccentric housing' acts in the same manner as a conventional wedge and sets the amount of 'kick' or 'dogleg' in the deviation process.

Reverse circulation drilling is a technology which has grown in application since first introduced in the early fifties. This process has developed over recent years to offer chip samples as opposed to lengths of core. The drilling mechanism is a pneumatic reciprocating piston commonly called a 'hammer' which drives a tungsten-steel drill bit. The term 'reverse' is used because air or fluid is blown down the rods and returned up the inner tube with the cuttings. Bits are large, often tri-cone and have round metal buttons which enable them to cut through unconsolidated or fractured material. The method is faster than conventional coring and does offer the option of switching to conventional coring once an ore body is reached. The drawback to this method is that while the process may be economical for the client, the initial investment for the drill company is higher in bits and in logistics. Heavier vehicles are used to convey water, auxiliary engines and maintenance equipment, and the heavier loads may adversely impact environmentally sensitive ground.

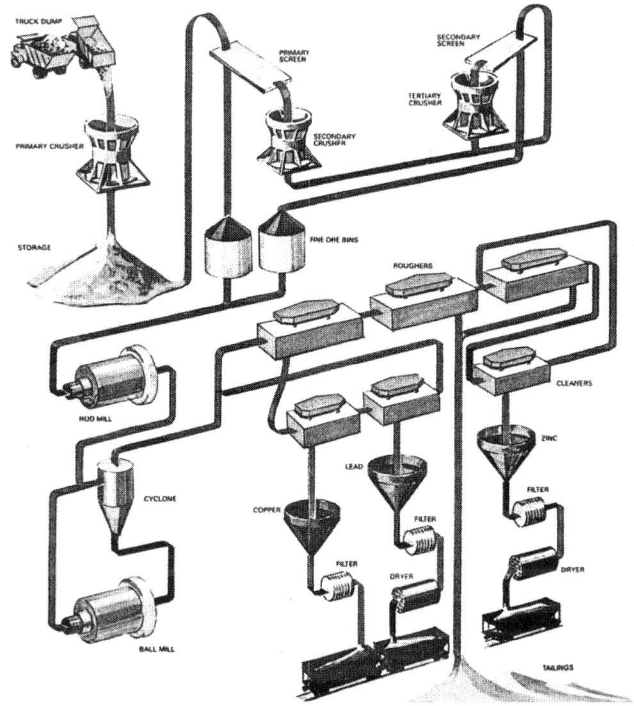

Ore processing flow chart of the former Mattagami Lake Mines, Ignace, Ontario

Diamond drilling technology and associated services have progressed because they comprise a highly competitive exploration component of the mining industry. An aggressive drilling association has closely monitored progress and acted as an advocate with government. Quite simply, it is now possible to drill a hole in any material anywhere but it costs money. Much still depends on the subjective nature of the driller's work. His most important skill is the ability to recognize what is happening at every stage of the process. At the company level, the emphasis now is on specialized drilling. Deep hole drilling, directional drilling and the ability to function in difficult environments and remote locations offer the most potential for growth. Limits of the technology such as current depth maximums will be overcome with investment in further research. The demand for drilling continues unabated, and despite the relentless pressure for technical advances and more sophisticated equipment, small Mom-and-Pop drilling outfits still find work.

An underground surveyor works at Hemlo.
Teck Cominco

Once drilling has located and outlined the ore body, it is generally found to be vein-type, massive or 'tabular,' meaning that it is like a flat sandwich of metals in the host rock. Companies have a variety of options available for extracting the ore. The mining methods, more sophisticated explosive techniques and mechanization, are far more extensive than they were at the close of World War II. Open pit mining is the first choice in mineral extraction because it is the least expensive operation. A big ore body close to surface will fit this method. Much planning is done in advance of mining. Rock mechanics determine the safe slope for the pit. Cost of mining the waste rock has to be determined so that does not exceed the ore value. With the advantage of vertical blast hole drilling, large shovels and huge trucks, production can be fast and a rapid return on investment achieved.

When the appropriate underground mining method is being considered, drilling results are combined with geological mapping, modelling and geomechanical characteristics of the host rock to assist in determining the correct way to go. Another development option is to take a look at existing properties that are similar to the one being investigated and see if these approaches would transfer easily to the current project. Once a mining method is chosen there are several considerations to examine to see if it meets project requirements. Safety costs and the mining schedule must be minimized. Recovery of 80% or greater of geological reserves must be achieved; waste rock dilution should be 20% or less. The degree of mechanization, automation and stope development must be considered. Trackless mining is an inexpensive means of transportation, readily able to meet changing conditions underground. The process must be flexible and adaptable in all areas.

Recalling ore bodies, the vein-type will be mined with small stoping underground; massive ore bodies require bulk mining with large stopes; and tabular ore bodies usually necessitate room-and-pillar mining.

There are four main bulk underground mining methods, listed here in order of increasing cost: block or panel caving is done by undercutting the ore zone and allowing the rock column to cave or fall downwards; blast hole or sublevel mining requires drilling long holes and then the ore is mucked or drawn off from below; sublevel caving takes place when the ore is drilled in rings and pulled or drawn off in successive lower lifts; room and pillar extraction involves working horizontally taking out a grid of rooms and leaving pillars to support the roof. Selective mining takes place when bulk operations are not feasible. Shrinkage is used with narrow veins which are sliced off in horizontal lifts. Cut and fill usually takes place by ramping down from a cross cut and then taking ore back in successive slices. The drift and fill means of taking ore is similar to cut and fill. In this method cuts in the rock are taken adjacent to one another. The drifts are then backfilled before continuing to the next area.

Forage Company drillers working in Quebec.

Golden Valley Mines

Automation systems are now winning acceptance. CVRD and Xstrata Nickel have pioneered such techniques. The Stobie Mine at the CRVD Sudbury nickel operation makes use of a sub level caving operation. Four long hole carbide drills are operated remotely, while load-haul-dump trucks can transport 3000 tons of broken rock a day on a 24-hour, seven-day-a-week schedule. Dumping and loading operations are carried out remotely by operators using cameras. Access to drills and the LHD vehicles takes place only when the machines are shut off. Vision sensors report the machines' positions to onboard automation equipment. Xtrata's Brunswick zinc mine started using tele-remote technology when it encountered weak ground. Automated underground mine trucks move between storage areas in the mine and are controlled from a central area. Sensors, onboard computers and cameras equip each truck. The system has significantly increased production rates.

Developing technology has enabled work on driverless open-pit haul trucks. The automation of underground jumbo drills is also useful as they remain stationary during operation. The main advantage of automation in mining is that it improves safety and removes the possibility of human error, lost time and human suffering through injuries.

An advertisement in *The Northern Miner* describes a mine in southwestern B.C. where changes in the mill have created a need for a graduate metallurgist. The job is to provide support in all milling areas resulting in an improvement in the overall economic viability of the mill. Early in the twentieth century when the ores of the Cobalt silver camp where found to be refractory or hard to process, the skills possessed by a modern metallurgist would have been much in demand. Today research groups such as the Applied Research and Technology Group at Trail B.C. and the Product Technology Centre in Mississauga, Ontario readily tackle problems in the metals processing field. Knowledge of the process of separating metals from their ore is much in demand in the mining industry.

Exploration samples give technicians the opportunity to predict the means of processing before a mine goes into production. Technicians use the processing history of metals and their alloys to determine the alloy strength, ductility, magnetic response and homogeneity of metals used in commercial applications. In the eighties an inexpensive method of recovering gold from low grade oxide deposits was developed called 'heap leaching.' The process is similar to the way water passes through ground coffee, dissolving the plant extract to produce the beverage. Broken ore heaped on a liner sheet made of thick polyethylene has dilute cyanide solution passed over it. In this trickle-down process the gold is dissolved. The gold-bearing or 'pregnant' solution is then run through a conventional Merrill-Crowe precious metal circuit.

The system developed in the early sixties in the United States to enable processing of refractory gold bearing sulphide materials is referred to as 'pressure oxidation' or 'autoclaving.' This relatively expensive metallurgical method involves the use of high temperatures and pressures to extract the metal. The most common extraction method uses cyanide to dissolve gold from finely crushed ore. What is known as the 'McArthur-Forrest process' converts the gold to water soluble aurocyanide ions. This method is subject to strict discharge guidelines due to the toxicity of the resulting effluent. The carbon-in-pulp or CIP process forms a slurry or 'pulp' of fine grains of ore with water added. Dosed with cyanide, it is agitated in tanks and absorbed by granulated carbon that is passed through the tanks to collect the gold, which is then separated from the carbon in the final tank.

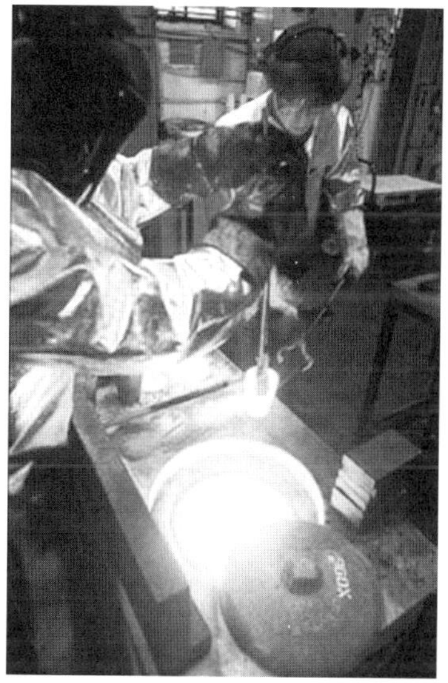

A refinery operator takes a sample from a molten gold deposit.
Royal Canadian Mint

When the ore body has been delineated by surface exploration, diamond drilling and sampling underground, a bulk sampling of the mineralized rock representative of the ore body not only determines metallurgical characteristics but also inspires investor confidence in the viability of the project.

In the case of bullion, mining companies send out their end product of silver and gold to be refined. The venerable Royal Canadian Mint is a highly profitable commercial crown corporation which has maintained its position as the leading refiner of gold in Canada since 1911. The process has changed little over the past few decades. Recently the Mint underwent much renovation but this progress is not evident in the long, high ceiling and rather dark refinery hall. Rough gold, which varies greatly in purity depending on the primary smelting at the mines, is first melted in an electric induction furnace so that uniform samples may be taken for assaying. Chlorine gas is then injected into the molten mass with the effect that the silver and base metals are bound in the mix but the injection is stopped before the gold is chemically bound. A thermal lance tests the molten mass to ensure that it has reached the required 2200 degrees Fahrenheit. Some of the base metals are picked up in the ventilator system and collected by an electric precipitator. What molten silver and base metals remain are skimmed off the top of the mix. The gold in the crucible assays

Gold is poured into a small mould for sampling.
Royal Canadian Mint

at 99.5 percent pure and further electrolytic refining brings the bullion to 99.99, the finest quality obtainable.

Mining engineer Rod Beauprie runs the Mint's refinery. His pride and joy is the new silver refinery which came on stream in 2006. For more than two decades the silver recuperated as a by-product of its gold refining was sent out to other refiners. As the price of the metal rose, it became advantageous for the Mint to commission a Swiss-based company to design and install a state-of-the-art silver refinery. The compact refinery is highly automated, operated almost entirely by computer. First step is a pre-melt to enable assaying. Molten silver is poured into a bath of cold water which produces impure silver grain. Next the grains are bathed in a nitrate-based solution. Copper and gold residual by-products are removed and the resulting .9999 fine silver flakes are cast into bars and other products. The compact refinery works away in a clean — almost sterile — environment with only one operator overseeing the process which currently produces two million troy ounces per year.

Unlike the United States, Canada does not have a bullion depository. All of the gold and silver in the high security vault belongs to customers or the Mint itself which uses bullion for high value coinage. The vault is a drab, functional storage place but the area comes alive in a warm yellow glow with the riches it contains. The Royal Canadian Mint does well by its master, the federal government and the mining industry it serves. The cover of its Annual Report has a one-word summary of activity: "Thriving."

Beating All the Odds

Successful Canadian Discoveries and Production

Remember what Champlain said: "Things go a lot better when you have a big outfit like Noranda backing you."

Advertisement aimed at prospectors during the Hemlo rush.

In 1988 *The Northern Miner* magazine gave a definition of a junior mining company: the typical outfit would not be a producer and would have no way of raising exploration funds except from the sale of treasury shares. Small size would naturally lead to 'junior' by definition. By contrast, senior firms derived their income by mining or other related business. Identifying mining successes during the period 1947 to 1981, the magazine listed seventy-four discoveries by seniors, twenty-eight by juniors and ten by senior and junior collaborations.

We have eminent scientist Dr. Donald Cranstone to thank for most of the papers in recent years on mineral discoveries in Canada. In 1988 he wrote in an article titled *Debunking the Myth*: "Junior exploration firms aren't the mine finders par excellence they're made out to be. The big guys found more mines before flow-through, and they did it cheaper than the little guys." He even debated the matter in print with veteran mining man and one-time PDAC president Fenton Scott. They did not arrive at a common understanding; the sticking point was in definitions of discoveries and their origins. Scott said that in any selection of excellence between individual and group initiative, history favoured the individual. They did agree on one thing: Canada's base metal reserves were running out. They were both incorrect on another point of agreement: they placed the timing of reserve exhaustion somewhere in the mid-nineties.

A ptarmigan pays a visit to the camp.
Diamonds North Resources

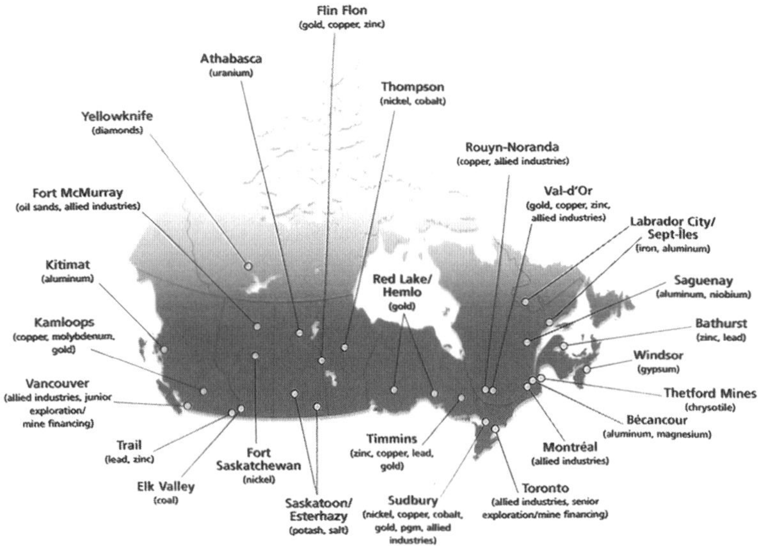

In the first forty years following World War II, various studies indicated that it was becoming more difficult to find new ore bodies. A highlight of the post war period was the exceptional rate of discovery from 1954 to 1956. During that time the uranium deposits at Elliot Lake, Ontario; the Thompson nickel deposits of Manitoba; the zinc, lead, copper and silver deposits at Matagami, Quebec, Manitouwadge in Ontario and Bathurst-Newcastle, New Brunswick; niobium deposits in Ontario and Quebec; and the copper and zinc deposits in the Flin Flon region of Manitoba were presented to a metals-hungry world. This was the time of the growth of airborne geophysical surveys. The only previously comparable time of great success in Canada was in the late nineteenth century with findings of nickel and copper. Some success may be attributed to luck — according to Stephen Leacock — in the way that luck plays a large part in success, as long as one works hard, and in the way of self-motivated prospectors in the bush. But, in reality, the majority of these deposits were found by trained geologists using the new scientific tools of metal detection.

There are some 190 producing mines and 40 operating smelters and refiners in more than 115 communities across Canada.
Mining Association of Canada

The American Mineral Information Institute estimates that each child born in the U.S.A. will need these minerals, metals and fuels in his or her lifetime:

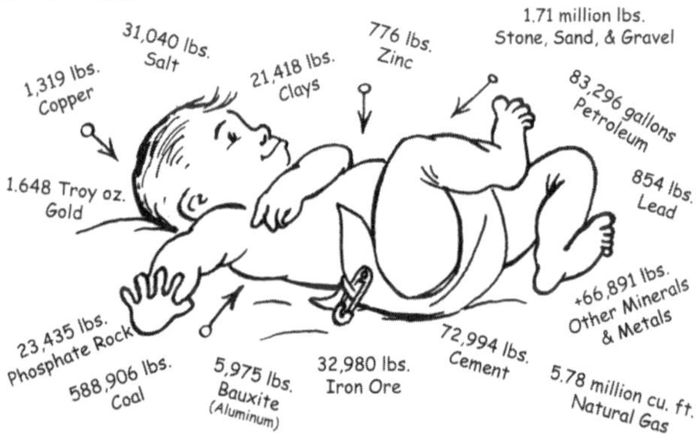

During those forty years of discovery the average time between discovery of an ore body and initial production was six years, half of which was spent in preparation for mining. Compare this with the discovery of the metals at Galore Creek in B.C. more than forty years ago where the deposit is only now coming to the development stage. Factors that also entered into the picture were the demand for metals, the economic conditions at the time, and the location and comparative difficulty expected in extracting the commodity. The cost of exploration rose sharply after 1953 and more slowly in later years. Nickel discoveries peaked in the mid-fifties and have declined since, only rising sharply many years later with the Voisey's Bay discovery. Copper enjoyed a peak discovery period from 1966 to 1971 due to the porphyry copper and copper-molybdenum deposits of B.C. Zinc had its turn from 1951 to 1956 in the Western Canada discoveries; a peak from 1963 to 1965 with Kidd Creek in Ontario and Cypress Anvil in the Yukon; and then another rise from 1972 to 1977 with a large deposit at Howard's Pass along the Yukon-Northwest Territory boundary. Lead followed the zinc pattern. As for molybdenum, the major discovery period was from 1960 to 1971 which was accomplished by a better understanding of the nature of the B.C. porphyry deposits. There were only two major molybdenum discoveries between 1975 and 1980. There were ten major silver discoveries from 1946 to 1981.

The big gold event in Ontario was the discovery of the Hollinger mine in 1910 and the related mines in the Porcupine and Kirkland Lake camps. We are aware of the rise in the price of gold in the late sixties, and yet from 1960 to 1962 there were five significant gold deposits located despite the fact that the yellow metal was not an attractive financial proposition for big investment at the time. Naturally, exploration rose with the climb in gold prices, but the Hemlo gold deposit discovery in 1981 was the biggest bonanza to take place in seventy years. The search for uranium in Canada since World War II has been governed by world demand. Once radium ceased to be the focus, and nuclear weapons and later power generation requirements took over, the demand for yellowcake soared. Although the first commercial reactors were developed in the fifties, the Canadian government price offers were too low for industry to bring known deposits into production. Only when the ante was upped to \$10.50 per pound did mining and exploration for the U_3O_8 began in earnest. Bancroft and Elliot Lake, Ontario were the hot spots, as well as the Beaverlodge area of northern Saskatchewan. Then exploration dried up in the face of limited demand and only began to show activity in the late sixties. At present Saskatchewan leads in world production and the price of yellowcake has skyrocketed, not the least influenced by flooding and a consequent stall in production at Cigar Lake, which would eventually prove to be Saskatchewan's biggest producer. As for other metals, the peak for tungsten occurred between 1969 and 1971 with deposits found in the Yukon and New Brunswick. Antimony peaked, enjoying the same discovery spurt in that same two-year period. There was a large copper-cobalt discovery at Windy Craggy in northwestern B.C. in 1981-1982, but that will never be mined as the whole region has been declared off limits as a provincial park.

One significant result of an examination of the 900 deposits discovered between 1946 and 1982 is that more than 140 of these were found in the last five years of that time frame. Significant discoveries are still made by prospectors; especially where glaciation has exposed deeply weathered rocks. However, the use of airborne- and ground-based electromagnetic, magnetic and radiometric geophysical methods have increasingly given geologists the power to more precisely locate minerals bodies of economic worth. Exploration costs, not surprisingly, have risen dramatically over the past sixty years; Cranstone has put the cost of finding a large ore body at between \$100 million and \$165 million as expressed in the year 2000 dollars. This is only an educated guess, but even with that scale of expediture, there is no guarantee of success.

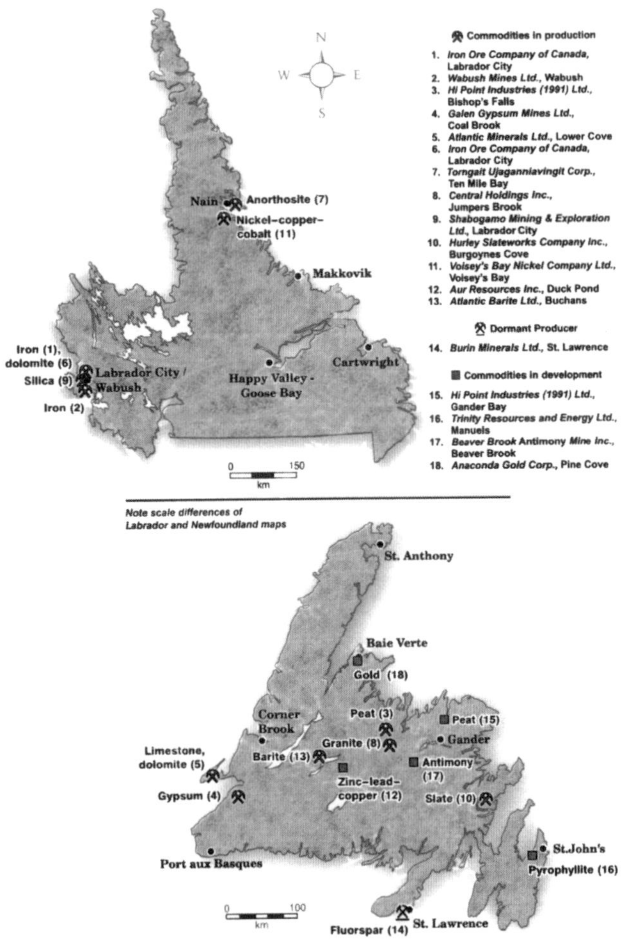

Producing mines and major deposits in Newfoundland
Newfoundland Natural Resources

A Primer on Mineral Deposit Types in Canada

Eight major mineral deposit groups are examined in this book; they comprise the metals, uranium and diamonds. Agricola, in his book *de Re Metallica*, (*Of Things Metallic*), wrote the first work on the subject, and ever since, geologists have been attempting to classify ore deposits into useful types. Once a useful classification scheme is devised, the characterizations that define each type of deposit become useful as exploration guides. Factors such as key host rock types, ages, forms and metal contents are all used to enable classification of each deposit into the correct group. Means of classification have evolved as knowledge of ore-forming processes has improved. The descriptive aspects and underlying scientific analysis of how each deposit type is formed is called a 'deposit model.' There are sixty-two deposit types listed in the bibliography of a book on the subject by Eckstrand and Sinclair. From this list, twenty-five have been major mineral producers in Canada.

For our purposes, all the sixty-two deposit types can be placed into one of three major groups or deposit types that fit into these groups. Stratabound deposits are those formed in or found exclusively in bedded rock. The strata may be sedimentary or volcanic but generally the deposit types were formed at the same time as the strata in which they occur. The second deposit types are those that were formed in intrusions or were deposited into adjacent rocks from fluid that emanated out of them. The third main group may be said to be enigmatic because they encompass a wide variety of vein and 'replacement' types which formed much later than their hosting rocks. Among some of these, the source of the metals remains unresolved.

Dr. Jim Franklin has kindly written introductions to each of the eight main commodity sections which are found at the beginning of each of these chapters.

Arnold Hoffman once voiced the industry truism: "Below ground, only the miner can shake hands with ore." This is so but, incredibly, above ground there are no easily-accessed sources for the details of the all the mines that have operated in Canada since the end of World War II. At least 700 metalliferous mines have operated in the country over the past six decades and now there are only seventy metals, diamond or uranium properties delivering their commodities in Canada. The author and his advisors felt that there should be some record of these mines. Therefore, output data has been researched and placed at the end of the book in a section titled: "Mines of Canada: 1946-2006." The statistics are incomplete and readers are urged to fill in gaps. Those with such knowledge are urged to contact John Harvey at *Harvey.information@hotmail.com* with information that may be added to these lists for inclusion in subsequent editions of this book.

Production statistics (tonnes mined and grade of the ore) are the only vital numbers about Canada's mining industry retained for posterity and, more importantly, that record the accomplishments of our mining communities and their people. Nowhere is there a single site where the country's complete mining statistics may be accessed for reference by our explorationists. Canadian federal, provincial and territorial governments, along with the various mining associations across the country, have been most negligent in failing to preserve this important facet of our mining history. The statistics provided are a modest start in that direction.

All That Glitters

Those Stratabound Deposits

Three types of stratabound deposits have played key roles in making Canada a leader in base metal production. They are volcanogenic massive sulphide, sedimentary exhalative deposits and Mississippi Valley-type deposits. There is a fourth type, stratabound iron, which will be mentioned later. The metals in all of these are contained in sulphide minerals, namely chalcopyrite (copper), sphalerite (zinc), and galena (lead); pyrite, the main iron sulphide mineral, is commonly the principal mineral of the first two types of deposits.

Volcanogenic massive sulphide deposits or (VMS) have provided copper, zinc, lead, silver and gold from every province in Canada except Alberta and P.E.I. Important discoveries of this deposit type have been made in the N.W.T., Nunavut, and the Yukon as well. Production from these will not take place until transportation routes have been provided for their concentrates. As their name implies, these deposits were formed in volcanically active areas, but only where volcanic eruptions occurred under great depths in the oceans. They are found in rocks of all ages from the very oldest greenstone belts of the Canadian Shield to some that are forming today off Canada's west coast, on the Juan de Fuca Ridge. The 'black smokers,' popularized in several 'Discovery' and 'National Geographic' features, are modern examples of these deposits actually forming. Examples of giant deposits of this type include the Kidd Creek deposit in Timmins, Ontario, one of the richest in the world, and the Brunswick #12 deposit, a giant among the many deposits of the Bathurst camp in New Brunswick. Many of the over 120 VMS deposits in Canada were discovered after the advent of airborne electromagnetic prospecting, because they are particularly responsive to this type of remote sensing.

The most volcanically active parts of all ocean basins, past and present, are at the areas where they are splitting apart, usually called 'oceanic rifts' or 'spreading centres.' Many of these are formed by splitting volcanic arcs, for example, in Japan and the Aleutians, whereas others may have formed in ancient mid-ocean ridges, similar to the Mid-Atlantic Ridge that extends from Iceland to the South Atlantic. In these rifts, very hot magma, melted from the upper mantle, rises towards the sea floor where it causes the volcanic strata below the sea floor to become unusually hot. When these rocks reach about 400 C, usually about two kilometres below the sea floor, the trapped sea water reacts with the rocks and leaches out most of their copper, zinc, lead, silver and gold, along with a lot of iron. This hot fluid rises quickly to the sea floor, usually along enormous fractures that bound these rifts. Upon reaching the sea floor they cool from near 400 C to about 2 C, causing almost all of the metal to precipitate. Massive sulphides build up as mounds on and replacements just below the sea floor. Leakage of this fluid from these mounds forms the 'black smoke' that scatters round the ocean floor, leaving a trail that enables explorationists to track down these deposits.

A second important type of stratabound in Canada is in the 'sedimentary exhalative' or 'SEDEX' deposit type. These formed in rifts that penetrate from the ocean floor into the continent, termed 'intracratonic rifts,' and are particularly abundant worldwide in mid-Proterozoic strata, which were deposited around 1300 to 1500 million years ago. These rift valleys were usually filled with shale and sillstone, commonly with some volcanic-related intrusive sills or a small amount of volcanic strata in the nearby sequence. These are giant cousins to the VMS deposits, as they are at least ten times as big as a typical VMS deposit. The fluids that make these deposits are generated in a similar manner to those for VMS, but form at a lower temperature, and usually contain only zinc and lead, some with silver. The giant Sullivan deposit near Kimberly B.C. is the best Canadian example of this type, although there are some excellent examples in the Howard's Pass area of the N.W.T. and the Tom and Jason deposits in the Yukon. The fluids that formed these deposits discharged into an anoxic sea floor, where reduced sulphur was abundant (usually from the decay of organic material). This sulphur trapped the metals and caused them to precipitate in thin but very laterally-extensive layers. Today this type of deposit is forming on the bottom of the Red Sea.

A third type of stratabound deposit, called Mississippi Valley-type (MVT) deposits, formed in carbonate sequences that were deposited in ancient shallow seas, one of which covered the North American continent during the Paleozoic Eon, about 450 million years ago. There are distant cousins of the SEDEX type of deposit that were formed from cool (-120 C) salty brines that travelled hundreds of kilometres through sandstone aquifers. Water from these aquifers percolated into paleokarst caverns only a few hundreds of metres below the land surface, formed by groundwater percolating down through limestone. These ancient caves and fault-related fractures were filled with organically-generated sour gas that was rich in H_2S. Lead and zinc precipitated from the dilute brine, slowly filling the caverns. Giant crystals of galena, sphalerite, calcite and quartz typify many of these. In Canada we have three major districts where these have been mined. The Pine Point District (76 million tonnes with 6.5% Zn, 2.9% Pb) just south of Great Slave Lake in the N.W.T. was the first, and ultimately the largest, Canadian district to be mined. Most notably, however, two mines operated for many years in the high Arctic: Nanisivik, at 73 N near Arctic Bay in northern Baffin Island; and Polaris, on Little Cornwallis Island. Both are in Nunavut and produced lead and zinc until recently. Polaris had the distinction of being one of the most northerly mines anywhere in the world, near latitude 75.4 N. There are several other examples of this deposit type in B.C. (Robb Lake, Monarch-Kicking Horse Pass), Nova Scotia (Gays River) and Newfoundland (Newfoundland Zinc).

There are several other stratabound deposit types that occur in Canada but they have not been extensively mined. One of these is called 'Red Bed Copper type,' somewhat of a misnomer since most of these actually occur in thin, black, organic-rich shale sandwiched between more oxidized red shale. They have been the mainstay of copper production since antiquity in Germany and Poland, and more recently in Michigan's Keweenaw Peninsula. Although there are good examples of these deposit types in Labrador and British Columbia, there is no production from them in Canada.

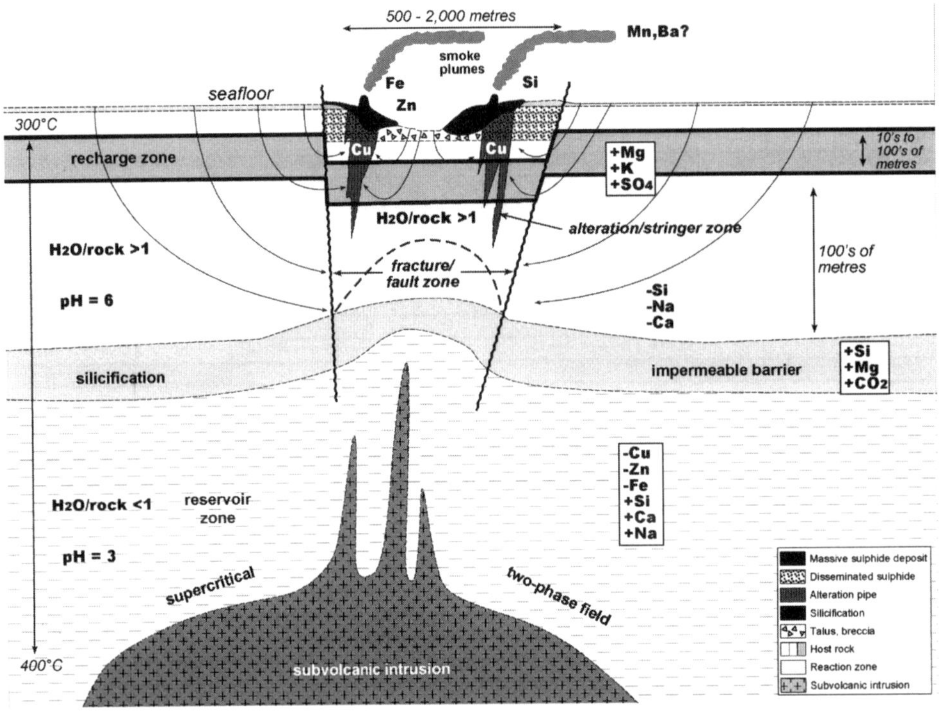
500 - 2,000 metres
Mn,Ba?
smoke plumes
Fe
Zn
Si
seafloor
300°C
Cu
Cu
recharge zone
+Mg
+K
+SO4
10's to 100's of metres
H2O/rock >1
alteration/stringer zone
H2O/rock >1
fracture/ fault zone
100's of metres
pH = 6
-Si
-Na
-Ca
silicification
impermeable barrier
+Si
+Mg
+CO2
-Cu
-Zn
-Fe
+Si
+Ca
+Na
H2O/rock <1
reservoir zone
pH = 3
supercritical
two-phase field
400°C
subvolcanic intrusion
Massive sulphide deposit
Disseminated sulphide
Alteration pipe
Silicification
Talus, breccia
Host rock
Reaction zone
Subvolcanic intrusion

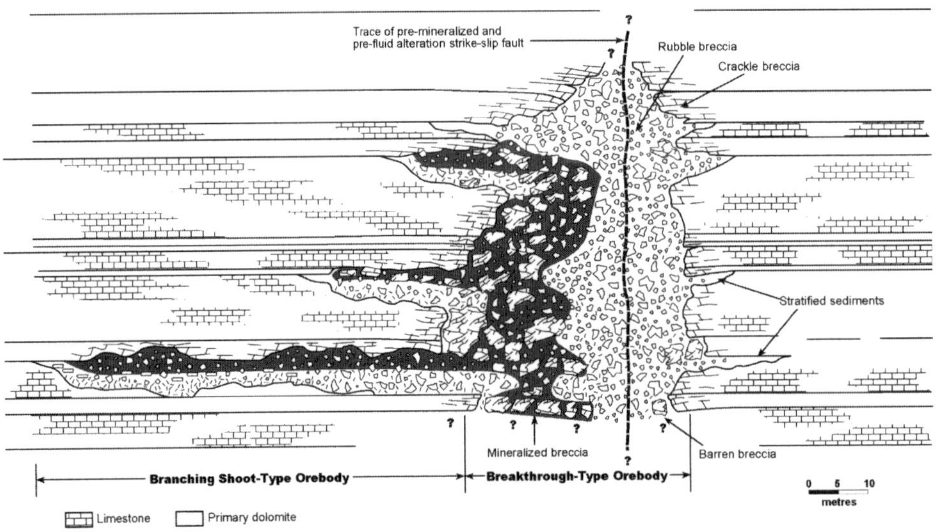
Trace of pre-mineralized and pre-fluid alteration strike-slip fault
Rubble breccia
Crackle breccia
Stratified sediments
Mineralized breccia
Barren breccia
Branching Shoot-Type Orebody
Breakthrough-Type Orebody
0 5 10
metres
Limestone
Primary dolomite

Those who search for copper, lead and zinc in large quantities know that their quarry will be found in massive sulphide deposits. One location is at Windy Craggy in northwest British Columbia. Visitors will observe it in its natural state as, regrettably, it will never be mined, being in a provincial park. The expression 'massive' does not refer to size but rather to the fact that the minerals in place are practically all sulphides.

Central Newfoundland was well-known for its once-rich Buchans and Daniel's Harbour mines, and the current AUR Resources' Duck Pond is the new mainstay of the industry in the area. Noranda started work on the property in 1973, and it took more than twenty years using all available techniques to outline the ore body, which was found to contain a large VMS mineralization consisting of a submarine deposit of sulphide lenses with pyrite, chalcopyrite, sphalerite and galena. AUR bought the property in 2002 to operate both as an open pit and as an underground mine. Over the estimated life of the mine, there were to be 4.1 million tonnes of reserves; the mine was truly polymetallic, having 3.3% copper, 5.7% zinc, 5 grams to the tonne gold and 0.9 grams to the tonne silver. All this would be put through the mill at 4,000 tonnes per day. Start-up was aided by the fact that AUR moved the mine and mill equipment from its Louvicourt mine near Val d'Or, which is now closed.

Although gone, the Buchans Mine is well worth remembering. Mineralization had been recognized in central Newfoundland for years, and in 1905 prospector Matty Mitchell prospected the site of interest and confirmed that it was visually rich. He received $2.50 for his work. He is remembered today in all Newfoundland government mining presentations. Technical problems in separation processes took years to resolve, so the mine that resulted did not

The Duck Pond Mine, Newfoundland
AUR Resources

begin operation until 1928. Over the next 56 years the VMS deposit gave up 16.2 million tonnes of ore. Today there is still much exploration activity in the area.

The Brunswick mining camp encompasses a volcanic outcropping at surface shaped like a four-leaf clover, 60 kilometres in diameter. The first VMS deposits were identified and drilled in 1938 at Orvan Brook. But it was not until 1953 and the discovery by the M. J. Boyle prospecting interests, which resulted in the Brunswick Mining and Smelting Company #6 Mine, that the area rose to prominence among the mining fraternity, resulting in a huge staking rush. Among world-class VMS deposits, this location had a number of 'firsts.' These included the area description in terms of a synergetic sea floor model, discovery by airborne magnetic survey, discovery by stream-silt geochemistry and the first Canadian heap-leach operation to recover gold in the deposit from surface oxidation. To date 45 deposits have been located in the area. Production from 12 deposits there, from 1957 to 1998 alone, was 12 million tonnes of zinc, copper and lead from an estimated total of 500 million tonnes of ore. The first mine to give up ore was the Heath Steele in 1957, but the star of the camp is the Brunswick #12 Mine, acquired by Noranda Mines, which has averaged 9,500 tonnes per day and in recent years has delivered grades of 3.49% lead, 8.81% zinc, and 0.34% copper. The mine is presently the largest underground zinc producer in the world but is scheduled for closure in the next few years.

Blue Note Metals is adding more mining to the Brunswick camp. The company acquired former producers Caribou and Restigouche mines from Breakwater Resources and reopened them in May 2007. These former operations, the deposits discovered by Anaconda in 1953, were at one time worked by Noranda but closed due to low metals prices and processing problems. The former is an underground mine and the latter, 50 kilometres west, is an open pit mine. Annual output has been projected to be 100,000 million pounds of zinc. The Caribou deposit reported a resource of 3.81 million tonnes grading 3.26% lead and 7.5% zinc, while Restigouche held 1.46 million tonnes grading 5.5% lead and about 7.1% zinc. The site was undergoing tailings remediation prior to the reopening.

The Horne Mine made Rouyn-Noranda and was one of the great 'elephant' copper gold properties in the world. Found by prospector Ed Horne who paddled into the remote area from New Liskeard, Ontario in 1921, the struggling new company got a boost in 1927 when gold magnate Noah Timmins of Hollinger loaned $2,000,000 to get the operation started. The ore was so rich that he was repaid within two years and the mine operated from 1928 to 1989. The deposit was VMS and today a huge smelting plant dominates the skyline around Lake Osisko. Recent studies have shown the geology to be very similar to the Kidd Creek mine in Timmins. Eventually Noranda — a name from Canada's North — wound up as owner. Over its long lifetime the Horne Mine produced a rich bonanza of metals. When the company closed the plant, there was great local concern for the economy in the area as it has been the mainstay of the two amalgamated cities. Noranda saved the day by refining and adapting its huge processing plant to become the largest, most advanced copper smelter in North America. About 15% of the material recycled contributes 150,000 tonnes of copper annually. They have a nice specialized business in recovering metals from items such as discarded computers, televisions and telephones, which are shipped to the twin cities from all over North America. Now Alexis Minerals and Xstrata are actively pursuing exploration of the west zone of the Horne ore body and other former producing properties in the area to locate more mineable resources. Chances of discovery are good as the properties under investigation are in-ground where there were once forty mines within sight of the great smoke stacks. The historic production of the nearby Quemont Mine produced a steady stream of smelter feed from 1944 to 1971.

The Duck Pond Mine, Newfoundland

AUR Resources

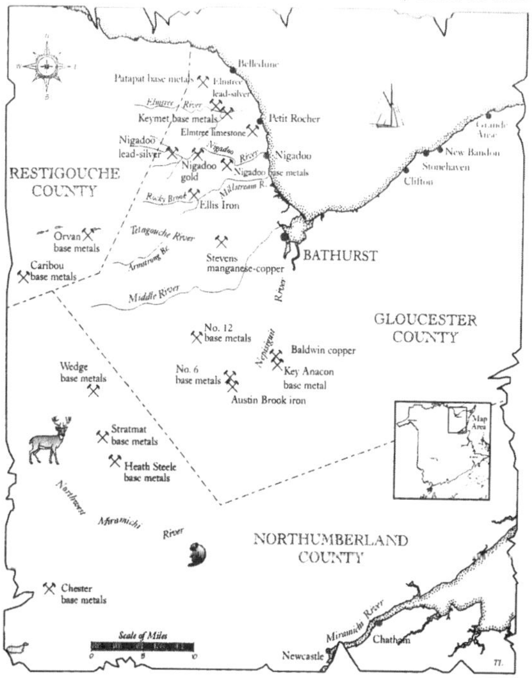

Summer must have been dry in New Brunswick when this map was drawn as the Nepisiquit and North West Miramachi Rivers do not appear.

Gwen Martin, Gesner's Dream

The La Ronde Mine is situated midway between Rouyn-Noranda and Val d'Or in the Cadillac-Bousquet region of the Abitibi greenstone belt.

The property was discovered by Noranda in the seventies and the miner found gold in massive and disseminated sulphide lenses. Agnico Eagle has three other gold properties in Quebec and the cash flow from La Ronde has paid for the projects under construction. The deep (2,200 metres) underground workings are cooled by an innovation, the first of its kind in Canada: water towers provide for refrigeration in summer months. The mine has been plagued by seismic events, so the underground workings are monitored by sensors which give warning of activity. Gold production comes 'free' because the mine also delivered 5 million ounces of silver, 77,000 tonnes of zinc and 7,500 pounds of copper, all of which pay for production. The reserves bring a smile to the faces of executives. The 5.3 million ounces waiting at depth will give rise to another mine — and that is another story.

Don McKinnon and Viola MacMillan study a claim map at the height of the Texas Gulf staking rush in Timmins, 1964.

The Matagami area of northwestern Quebec in the Abitibi greenstone belt has eighteen known VMS deposits, most of which were discovered in the fifties by airborne geophysics. The location is the southern portion of the James Bay road. The name Matagami is Cree for "where the waters meet." The largest of these properties was the Mattagami Lake Mine, discovered in 1956 and mined from 1963 to 1958. The Selbaie Mine, 150 kilometres west of Matagami, was owned by BHP Billiton. Exploration of the area was hampered at first by heavy overburden but the deposit was eventually outlined by reverse circulation drilling. There were several smaller properties including the Poirier Mine discovered by Rio Algom in 1959 and brought into production in 1965. The area is still open to further discoveries. In 1999 a new EM geophysical exploration system was flown over the Matagami camp, which led to the discovery of the three Perseverance deposits in 2000, handily located for processing at the Lac Matagami mill six kilometres away. Xstrata has announced that the deposit has a measured and indicated resource of 5.12 million tonnes of 15.82%

zinc, 1.24% copper, 29 grams per tonne gold and 0.38 grams per tonne silver. It is believed the mine will have a life of five years and produce 228,000 tonnes of zinc concentrate annually.

Breakwater Resources opened its Langois Mine at Lebel-sur-Quevillon, 213 kilometres north of Val d'Or in mid 2007. The recommissioned mine has a measured and indicated resource of 5 million tonnes averaging 11.1% grade zinc, 0.8% grade copper plus 54 grams per tonne silver and 0.1 grams per tonne gold.

Timmins is the smallest city in Northern Ontario but it has been known since 1909 as one of the great gold camps of Canada. Half a century later many of the big mines were closing. Salvation for the local economy came in the actions of Texas Gulf Sulphur, which was using the then relatively new science of geophysics, crisscrossing the country looking for sulphide deposits. A strong anomaly was located near the city in 1959 and more exploration work resulted in the discovery of one of the largest VMS deposits in the world. The announcement of the discovery in 1963 saw a huge staking rush in the area, which over the years has been the property of Falconbridge and now belonged to Xstrata. The operation began a year later as an open pit and later developed three shafts or mines in close proximity. An ambitious project in recent years has seen the ore body followed down currently to the 9,600-foot level, a trend which will see production until at least 2013. The metallurgical plant at Hoyle produces copper and zinc concentrates from its mill, copper blister from the refinery, as well as silver, indium and cadmium by-products. Total production from the mine to the close of 2006 has been 140 million tonnes grading 6.5% zinc, 0.26% lead, 2.4% copper and 2.39 ounces per tonne silver.

Xstrata also has a small nickel-copper deposit at the Montcalm property 70 kilometres northwest of Timmins. Discovered in 1976 by the Montcalm Syndicate, the ground was obtained by Xstrata predecessor Falconbridge in the mid-nineties. The mine was put into production in late 2004 and was predicted to have a life of 8.5 years. Production through its ramp and portal access creates a small footprint as the ore is processed at the Kidd concentrator. The copper concentrate is treated at the Kidd Metallurgical site and the nickel concentrate is processed to nickel matte in Sudbury. The average grade is 1.46% nickel and 0.7%

copper with an annual production of 9,000 tonnes nickel ore and 5,000 tonnes copper.

Today Ignace, a small village located northwest of Thunder Bay on Highway 17, markets itself as an outdoor tourism spot. There is no mention of the rich mining history of the area or the busy times that developed for the community after late October 1968. This is when Noranda affiliate Mattagami Lake Mines began exploring an area located in the Wabigoon Belt on the east side of Highway 599 that runs north to Savant Lake. Led by Chief Exploration Geologist John Harvey, the company would devote more than four years to developing a massive sulphide deposit beneath an area of 118 feet of overburden with no significant outcrops hinting at the bonanza that lay in the location. Ignace was the supply and service hub for the project which eventually, after airborne geophysics and 74,701 feet of diamond drilling, outlined a rich ore body.

First exposure of the Kidd Creek ore body

Brian Hester

The discovery brought a flurry of staking activity and other mines developed in the area. Company president W.S. Row told the *Northern Miner* that with 568 claims, the geology team would not suffer from lack of work. Apart from the normal problems that plague mine finders, negotiations had to be undertaken with Abitibi Pulp and Paper which held blocks of land as timber rights in fee simple that originated from the building of the railway westward during the nineteenth century by the CPR. Most vexing were the antics of a swarm of claim jumpers who descended like vultures on the project with ploys such as accusations of improper staking and other products of their imaginations. The company solved the problem by making a token small payment to one of the fraternity and left the rest to fight over these meager spoils. The Mattabi open pit, Lyon Lake, Creek Zone and part ownership in the Sturgeon Lake Mine proved winners and subsequent mining came in fast. John Harvey, who became Noranda's president of exploration in 1982, continued with the aim of obtaining more ore supply. His work resulted in the acquisition for Noranda of the much greater plum of the Golden Giant gold deposit at Hemlo.

The Flin Flon area of Manitoba is located north of The Pas and straddles the Manitoba-Saskatchewan border. Minerals were discovered in the early part of the last century, but when large VMS deposits of zinc and copper hosted in volcanic rock were located, the Hudson Bay Mining and Smelting Company established Flin Flon in 1927. The area has consistently produced ore since 1930; 40 million tonnes have been found in the past twelve years. There was a rapid impetus for mine discovery in the fifties and early sixties aided by the widely adopted technique of Canadian Mining Hall of Fame member Albert Koffman, who advocated saturation prospecting, geophysics and diamond drilling in the area.

The Kidd Creek mine has the longest underground ramp in the world.

Xstrata

The Flin Flon Mine operated until 1992 and production from that area now comes from HBMS 777 and Trout Lake Mines. Several satellite mines have produced in the vicinity over the years. In 1960 the Chisel Mine opened at Snow Lake, 120 kilometres east of Flin Flon, and eight mines have developed there since that time. The Hudson Bay Mining and Smelting Company is now the third largest producer of copper and zinc in Canada and is now under new ownership as HudBay Minerals. Annual production from all the company mines in Manitoba is 114,700 tonnes of zinc metal, 43,200 tonnes of zinc oxide and 86,300 tonnes of copper metal. The latest prospect is the Bur deposit near Snow Lake.

The Myra Falls Mine on Vancouver Island, ninety kilometres from Campbell River, offers excellent proof that mining can be done in an environmentally sensitive area. Located in Strathcona-Westmin, the oldest provincial park in the province, it presents no ecological bar to tourists who often take the mine tour. Operating since 1966, this complex VMS deposit is now owned by Breakwater Resources. Originally an open pit project, it now has two underground mines. The modern concentrator has an annual 1.4 million tonne capacity. From 548,225 tonnes milled in the first nine months of 2006, concentrates obtained were 27,057 tonnes of zinc at 5.7 % and 4,006 tonnes of copper at

1.0%. Silver production at 50 grams to the tonne was 685,074 ounces and the gold at 1.6 grams came to 16,777 ounces.

Pine Point Mine was located on the south shore of Great Slave Lake. The Mississippi Valley-type lead and zinc deposit was the subject of attention from Klondike miners who thought silver would be found there. The area was explored for six decades before being developed and opened by Cominco in 1964. A rail line to the site from Roma, Alberta, aided in construction and the plant produced until 1987. The mine was one of the largest and most profitable of its type in Canada. When the property closed, the adjacent settlement, which housed 1,200 people, was demolished and the land reclaimed. A junior, Tamerlane Ventures, now has the property and is exploring it further.

Two great mines were pioneers in the high Arctic and their base metals, Mississippi Valley-type deposits, each lasted more than twenty years. Teck Cominco's Polaris Mine, on Little Cornwallis Island, Nunavut, was the most northerly base metal mine in the world. The lead zinc deposit was discovered in 1960 and in operation from 1981 to 2002. The entire surface plant was built in Montreal, then shipped in by barge and anchored there along with the camp and all facilities under one roof. At Breakwater Resources' Nanisivik Mine on Baffin Island in Nunavut, the mine was innovative in that primary stopes were filled with frozen backfill to preserve the integrity of the permafrost. The company pioneered training of Inuit residents and developed many large dome-shaped prefabricated buildings to provide accommodations, including a swimming pool.

A small Mississippi Valley-type deposit at Gay's River near Truro, Nova Scotia, should come into production in 2007. The Scotia Mine had operated periodically with difficulty due to underground water problems and the last owner, HudBay Minerals, sold it to Acadian Gold. The property, until now on standby, has 4.6 million tonnes averaging 3.6% grade zinc and 1.7% grade lead with renewed production contemplated.

Across the country, the Yukon may well have the next major VMS deposit in Yukon Zinc's Wolverine deposit. Located on the original 1942 railway survey route northwest of Watson Lake, the find may be truly said to be polymetallic. Not only does it hold reserves and resources containing 1.7 billion pounds of zinc, it also has lead, copper, silver and gold. Construction of the mine began in 2007. Whitehorse Copper, a copper skarn deposit, was mined by Hudson Bay from 1987 to 1981 and produced 3.1 million tonnes averaging 0.95% copper. Sherwood Copper is a promising new property in the Yukon, which will deliver both gold and copper. Only discovered in 2006, the Minto project is situated northwest of Carmacks. The geology is quite complicated and the company states that there are no other deposits analogous to it in the world. Fast-tracking cooperation by the territorial government has meant the operation was ready to commence in mid 2007.

Massive sulphide lenses found in local basins on the sea floor are sedimentary exhalative or SEDEX deposits and are many times bigger than the VMS deposits. They are found in western Canada and there have been two large operations mining such deposits. One is the giant Sullivan Mine at Kimberley, B.C., which was discovered in 1892 by Pat Sullivan and partners, acquired by Cominco in 1909, and bought outright in 1913 for $200,000. The mine, consisting of a complex body of sulphide ores, closed in 2001 after producing more than $20 billion in lead, zinc and silver. This was from 150.5 million tonnes averaging 5.81% zinc, 6.12% lead and 61.56 ounces per tonne of silver. But the story does not end there. Visitors who come to vacation around Kimberley and view the museum about the Sullivan Mine might notice much mining exploration activity. Sedex Mining and Stikine Gold have both done deep drilling on the Sullivan fault. The other SEDEX deposit is the Faro Mine in south central Yukon. Discovered by prospector Al Kulan in 1953, the property was developed by Cypress Anvil and went into operation in 1969. For many years it was the largest lead and zinc open pit mine in Canada. There was a shutdown for a while in the eighties and the mine produced again under Curragh Resources, finally closing in 1992. Since 2003, First Nations have worked with the Yukon government on reclamation. Curragh Resources operated two other smaller mines near Faro called Vangorda and Grum — unkindly called "Grim" by some observers — between 1991 and 1998 with grades of lead and zinc similar to those of the Faro Mine.

Hudson Bay Mining Co. property, Flin Flon, Manitoba
Sask. Industry

Cementation Company workers pose after completing their underground development work at Kidd Creek.
G. Oxby

Many observers of exploration in Canada assert that any new discoveries will take place in the Arctic, on mountains or in other hard-to-reach places. As this book edged towards publication, a junior company called NorOnt Resources located a most promising nickel-copper and precious metals discovery in a lonely spot at its Double Eagle property. The location is 170 kilometres due west of De Beers' Victor Mine off the Attawapiskat River in the McFaulds Lake area of Northwestern Ontario. Some other juniors are also in the area; investors, as ever, have rushed to get in on the action.

Iron

Iron rusts from disuse

Leonardo da Vinci

Canada's iron production has come from two types of stratiform deposits known as the Lake Superior type and the Algoma type. It is a measure of the importance of these big producers to the North American economy that they are named after these districts. The difference between these two types of iron deposits and the base metals is that they extend for tens to hundreds of kilometres as thick, iron-rich sedimentary beds. The parts of these beds that are economic are defined only by the quality of the iron minerals. By far the more important of the two types are the giant Lake Superior deposits of the Labrador Trough; the ore type is commonly known as 'taconite.' Processed iron from these sources serves the steel mills of both Canada and the U.S. The smaller Algoma-type deposits have been important producers primarily for the Canadian steel industry. Lake Superior-type deposits formed in large rifts that probably came together at the margins of continents during the start of continental break-up. The iron and silica-rich fluids seem not to have had as focussed a discharge mechanism as the VMS and SEDEX deposits, but these more passive discharging fluids caused water in the rift basins to become highly saturated with iron and silica. Recent research has clearly emphasized the importance of ancient bacteria as a mechanism for forming the iron minerals together with biogenic chert, probably through the creation of oxygen and changing acidity in the basinal water. Iron minerals rained down from iron-saturated ocean water to make alternating iron-rich and chert-rich beds. It is a curious fact that many of these types of deposits worldwide were formed at a time in the earth's history when the atmosphere was becoming more oxidizing. The combination of continental break-up and oxygen generation, plus the onset of prolific bacterial development in the early Proterozoic period, seems to have favoured their formation. Other 'catastrophic' oxidation events, such as the melting of worldwide ice cover during the 'snowball earth phenomena' may also have engendered the formation of these deposits in younger (late Proterozoic) strata in Canada's cordilleran region (e.g. the Redstone deposits in B.C.). Iron oxides, magnetite and hematite, are the economic minerals of interest. Magnetite is easily separated using magnetic separators, and hematite is separated using floatation methods; both these methods produced a pea-sized, iron-rich pellet that is highly sought-after by the world's steel mills.

The Algoma-style iron formations are more closely associated with volcanism, but otherwise similar to the Lake Superior deposits. They tend to be thinner units, some of which are closely associated with VMS deposits, but most seem to have formed nearer the margins of volcanic arcs. Otherwise they contain the same minerals as the Lake Superior type, but in addition, may contain iron-rich carbonate and sulphide materials. Production from these came mainly from Ontario; the Michipicoten iron ranges near Wawa, the Sherman Mine at Temagami, and the Atikokan area mines represent carbonate, oxide and sulphide types respectively. The latter could only be produced because strong oxidation converted them to hematite during a period of global warming.

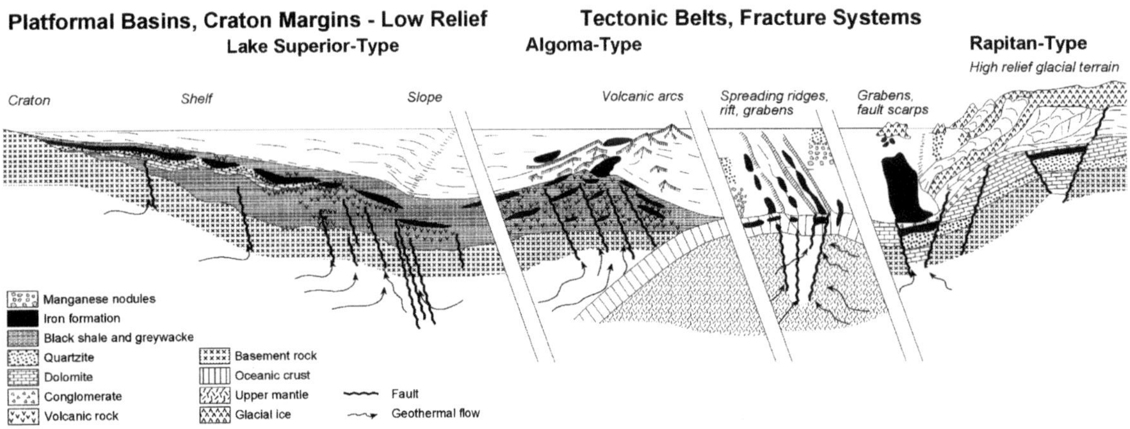

They do not mine iron, the main ingredient in the production of steel, in Ontario any more. Only large deposits that can be economically delivered are considered for mining as there is lots of competition in this particular commodity market around the world. The Marmora discovery was made in 1955 utilizing airborne geophysics to detect an anomaly. The resulting open pit Marmaraton Mine, operated for

Bethlehem Steel, milled 28 million tonnes at a rich 40% grade before closing in 1979. Today, the giant, partially water-filled pit is a tourist attraction. The low grade siderite ore at the Algoma Ore Properties ground at Wawa was discovered in the early part of the twentieth century and the plant became a subsidiary of Algoma Steel in Sault Ste. Marie. The Helen Mine had both open pit and underground operations, the later McLeod Mine was underground and there were also three other pits. Concentrate was shipped from Michipicoten Harbour on company ore carriers on Lake Superior to the Sault. The deposit was mined out and closed in 1998. The Adams Mine in Kirkland Lake and the Sherman Mine at Temagami, both of which opened in the early sixties, were located in banded iron formations, with the iron in magnetite, operated for a quarter century and were only closed due to the pressure of foreign competition. The Sherman produced over 84 million tonnes of ore which after concentration came down to 22,244,212 tonnes of pellets. The Adams returned 30, 275,000 tonnes of concentrates. Mines in South America could produce iron cheaper and Dofasco closed the two active properties, despite the fact that the Sherman pits had five years of reserves and the Adams Mine had twelve years of life in it at closure in 1990. (The City of Toronto later unsuccessfully proposed to use two pits at this location to dump the city's garbage.)

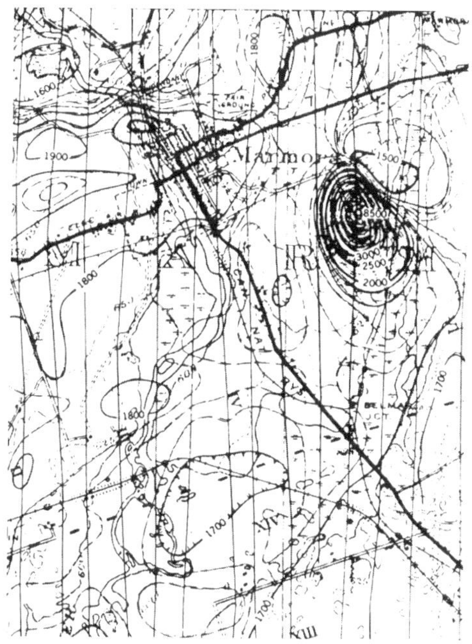

One of the first aeromagnetic discoveries in Canada was made at Marmora, Ontario. Vertical lines indicate the flight pattern, while contour lines show the iron concentration.
Morris Zaslow

Atikokan is an Ojibway word meaning "caribou bones." The area came alive due to the discovery of a large deposit of iron and the need for the metal during World War II. The presence of good value hematite and the strong possibility that the iron would be under Steep Rock Lake was known in 1897, but it was not until the late thirties when the work of prospector Julian Cross was backed up by Joe Errington's money that the deposit was confirmed by detailed exploration. The demand for iron for the war effort accelerated development. Today we are used to mega mining projects, but for the time when Steep Rock was being developed, diverting the Seine River around the lake and then pumping it dry was a tremendous engineering feat. By 1944 more than

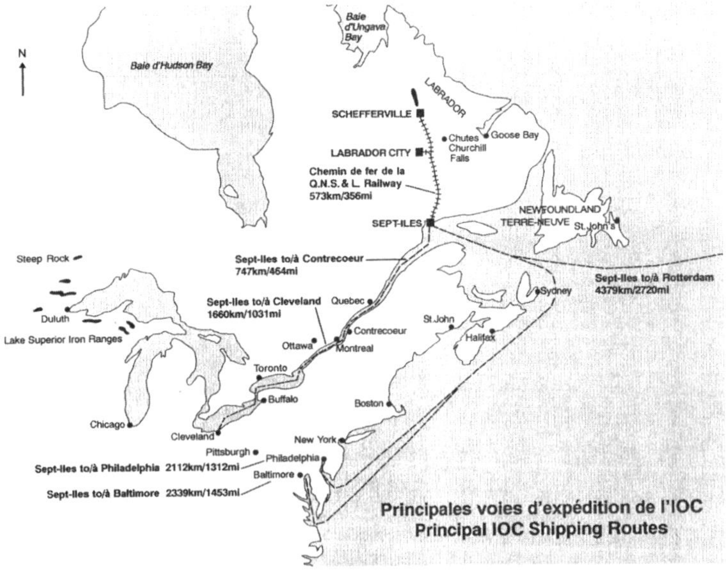

70 billion gallons of water had been removed, then a 400 feet depth of glacial silt and boulder clay trucked away and the first pit commenced operation. From 1944 to 1979 the original mine plus the Caland property, which began producing in 1966, delivered a combined total of almost 100 million tonnes of ore before the mines closed.

Iron mining continues in eastern Canada. Newfoundland mined the metal from 1894 to 1966 at the Wabana operation on Bell Island in Conception Bay, near to St. John's. In its time the two surface and four submarine mines extending a third of a mile under sea produced 78,989,419 tonnes for Sidney, Nova Scotia, mills and also for export. The mainly sedimentary-based hematite ore created some marketing problems due to high phosphorus content but in the end the less expensive Labrador properties put an end to Bell Island mines and visitors are reminded of the heroic endeavour with tours that include a part of the submarine workings under the Atlantic Ocean.

Canada is now ninth largest iron ore producer in the world and fifth largest exporter of the metal. Three mines in Labrador and Quebec ensure this position plus some magnetite recovery in British Columbia. Two of these properties, the Iron Ore Company of Canada and Wabush Mines, are located in Labrador West. Gwen Martin in her book *Gesner's Dream* (Canadian Institute of Mining, 2003) rightly calls the Labrador iron trough a treasure trove of ore. The Geological Survey of Canada found the hematite deposits in 1895 and the land was staked and examined by various companies over the years but it was not until 1952 that the present company was incorporated. After a huge development project the first ore was shipped from the

The iron concentrate train travels 563 km. from Labrador City to Sept Isles.

Schefferville-Knob Lake area, despite an almost complete lack of infrastructure and transportation routes in the area. The harsh nature of the vast landscape with mountains, lakes, tundra and muskeg did not deter the company planners. Over a four year period a workforce of 6,000 built two town sites, laid track and constructed the Quebec North Shore and Labrador Railway, 573 kilometres long, linking the mining area to a shipping terminal at Sept-Isles, and during the same period constructed two hydro-electric plants and built the terminal capable of handling 10 million tonnes of annual ore production. Construction and upgrading has continued to the present, including a huge concentrator project at the terminal base.

The Ship Sprint takes on iron concentrate at Sept Isles.

After twenty-eight years the Schefferville operation was closed in 1982, but there was plenty more iron to be mined. The Labrador City area had produced since 1962 and continues its huge output today. The capacity is an annual 17.1 million tonnes which is used to produce 10.2 million tonnes of pellets. An automatic electric train takes ore to the concentrator, and, after pelletizing, the product runs to Sept-Isles in up to 210 cars at a time. By 1996 the company had produced one billion tonnes of iron ore pellets. Half a century of production was achieved in 2005 and the following year a record one-day shipment of 46,454 tonnes of pellets was sent for export although the average annual range is 35-38 million tonnes of iron at a 40% grade. Areas no longer mined are reclaimed under the Iron Ore Company's Tailings to Biodiversity program by which thousands of hectares are converted into wetlands. Rio Tinto owns 59% of the massive undertaking which has reserves for at least fifty years. Dofasco and Bethlehem Steel each have a piece of the company and a relatively new shareholder is the Mitsubishi Company of Japan.

The Quebec Cartier Mining Company operations are in the Côte Nord region, part of the East Grenville Province of Quebec. Normanville Township hosts the property and the town of Fermont is 16 kilometres east of the mining complex. From the town, the mine appears like a mountain of rock off in the distance. Metamorphosed iron formations were the impetus for the Lac Jeannine Mine which opened in 1961; after being mined out, it was closed. Mont Wright Mine opened in

Geologists working on Irony Mountain, west of Schaefferville, Labrador
New Millenium Capital

1974, its four pits mining an average of 40 million tonnes a year. There is a road out to Baie Comeau, but the ore goes via train 416 kilometres to the pellet plant at Port Cartier on the St. Lawrence River where 13.5 million tonnes are produced annually for mostly export shipment. The grade of the hematite and specular hematite is only 30-37%, so Quebec Cartier has to be very efficient and look for economies to maintain its output.

The town of Wabush in the Wapusakatte Mountains of Labrador chose an appropriate motto when it incorporated in 1967. "From the Earth We Prosper" is indeed appropriate because the Wabush Iron Scully Mine is the main reason for the existence of the community. A much smaller cousin of the Iron Ore Company of Canada operations, the Scully Mine produced 5.3 million tonnes of pellets in 2006. The ore is specular hematite which is non-magnetic and has a manganese content which complicates processing. The operation, which started mining in 1968, is operated by Cleveland-Cliffs, which is also part owner with Stelco and Dofasco. The concentrate is shipped by rail to the pelletizing plant at Point Noire on the Seaway via a link with the Quebec North Shore and Labrador railway. So far, the mine is the only one of the three producers with a looming closure date. The ore is predicted to run out in 2013 unless more economic iron can be found.

Labrador City has a neat layout.
Cleveland Cliffs

A most promising iron ore venture is being developed by New Millennium Capital along the Millennium Iron Range in northern Quebec and Labrador. The LabMag and KeMag properties comprise a vast magnetic taconite deposit at Howells River Labrador about 30 kilometres northwest of Schefferville, Quebec. The mine life is currently estimated to produce 15 million tonnes a year for at least 65 years. Construction is due to start in 2010. Consolidated Thompson is developing its Bloom Lake mine which should produce 5 million tonnes of 66% grade iron in 2008.

Wabush Mines, Labrador
Cleveland Cliffs

Although Eastern Canada has held sway over iron production in Canada, there is a strong possibility in the not-too-distant future that another — to some, unlikely — location for iron production may deliver the metal. The deposit is at Mary's River, 160 kilometres south of Pond Inlet in south central Baffin Island, Nunavut. Well-known geologist Murray Watts located the prospect in 1962. Since then, further examination has indicated five high grade deposits grading a huge 68% iron. The massive specular hematite body is estimated to offer 35 years of mining at a million tonnes annually and has the advantage of being high grade 'lump' material, so that any miner would merely have to dig, crush and ship for delivery as direct blast furnace feed. Baffinland Iron Mines has to overcome infrastructure and transportation costs after its feasibility study is done. The place needs a 100 kilometre rail line, a deep sea port and ice breaker support for Arctic-going ships. Financing the $1.5 billion project might not be too big a hurdle as high world prices and strong demand, particularly from China, would give the company some strong encouragement.

Sept Isles dock terminal
Iron Ore Co. of Canada

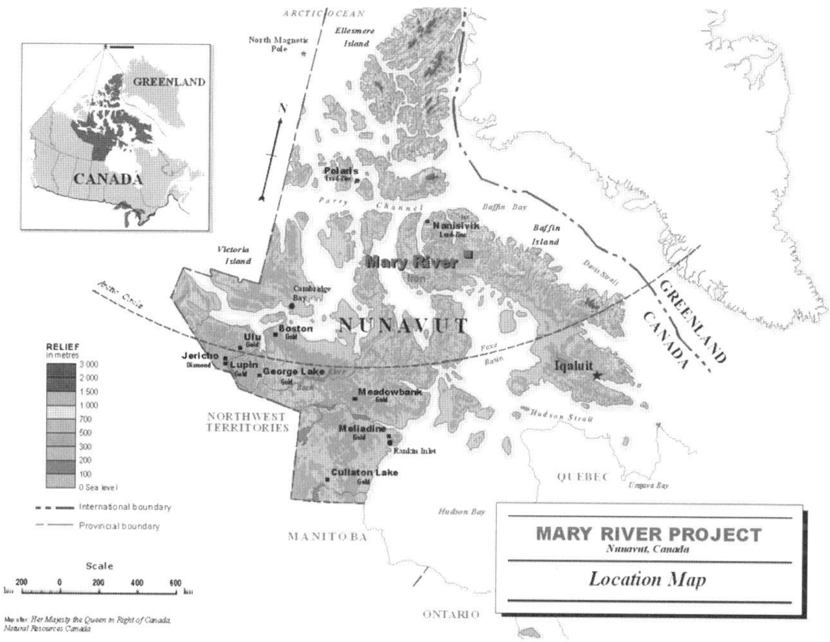

The Mary River project is on Baffin Island.
Baffinland Iron

The World Wants Yellowcake

Uranium

Some recent work... leads me to expect that the element uranium may be turned into a new and important source of energy...

Albert Einstein, Letter to President Franklin Roosevelt, October 11, 1939

The demand for uranium began with the development of the atomic bomb; uranium for the weapons used during World War II came from Canada, and the Cold War-sparked weapons race during the fifties and sixties created a short-lived demand. During the same period, Canada and other developed countries began to research and construct nuclear power plants which helped to sustain demand through the late sixties and early seventies.

Termination of the Cold War and release of uranium stockpiles for peaceful purposes caused a slump in prices, but in recent years, driven by the enormous demand for power by China and India, the market demand for uranium has increased remarkably, and the hunt is on for new resources.

Although the first production of uranium came from small vein deposits in the Great Bear Lake area, and also from granitic pegmatite near Bancroft, Ontario, it was the need for uranium to support Ontario's growing nuclear power industry that prompted the discovery of the Elliot Lake deposits.

Canada became the world's leading producer of uranium from two main deposit types: the Elliot Lake 'paleoplacer' type and the Athabasca 'unconformity' type. Both are stratabound, but their similarity to the base metals and iron deposits discussed above ends there. Historic and strategically important uranium production came from a third type, vein deposits, and a fourth type, pegmatite deposits. Elsewhere in the world, a fifth deposit type, sandstone or 'roll front' type, and a sixth, granite-hosted deposits, are important, but thus far none have been produced in Canada.

The Elliot Lake type deposits all occur at the base of an early Proterozoic (about 2300 million years ago) sequence of mainly sedimentary rocks that covered much of the eastern Precambrian Shield in Ontario and western Quebec at that time. These rocks were deposited at the end of a major period of glaciation, and in many respects they resemble the very recent (in geological terms, only 8,000-10,000 years old) glacial sediment that covers much of eastern, central and northern Canada today. They formed as lake sediments, glacial stream deposits, outwash from retreating glaciers and eskers. The source of the clay, sand and boulders that comprise the various stratigraphic units was the adjacent Archean landmass, which probably looked much as it does today: large areas of granite, surrounding crescent-shaped ancient volcanic terrains, the 'greenstone belts' that host the VMS and vein gold deposits mentioned elsewhere. Glaciation during the early Proterozoic seems to have been worldwide (an early snowball earth perhaps) and caused massive erosion of the Archean terrains. One important difference between recent glacial events and the one in the early Proterozoic was the very low oxygen levels in the earth's atmosphere at the time. Oxygen only became abundant at the end of the Huronian period.

The Elliot Lake deposits formed at the beginning of the formation of the Elliot Lake sequence, by the accumulation of heavy resistant minerals that were weathered from and washed out of the adjacent Archean highlands, and transported in fast-moving streams towards the south. The principal uranium mineral, uraninite, occurs in very small amounts in most granite bodies; along with quartz and pyrite the uraninite was washed into glacial streams, where the heavy minerals worked their way to the bottom through an agitation process that is much like the process that formed the Yukon's placer gold deposits and the swirling, agitated motion used by gold panners. The first discoverer of these deposits, Franc Jobin, recognized the results of this process at what became the Pronto Mine and traced these ancient stream channels back through the folded Huronian rock. This knowledge was used to guide geologists to discover all the deposits in the Elliot Lake camp. These ancient stream channels are composed of large quartz boulders with the fine, but much heavier uraninite (and its sister mineral, brannerite). Geologists mapped out the old streams and miners followed them. Typical deposits contain 0.05% to 0.1% U_3O_8, but they are huge; the 'pay zone' at Quirk Lake extended for 10 kilometres long and 3.5 kilometres wide and contained 200 million tonnes of ore. By the way, a similar process of paleoplacer accumulation of heavy metal may well account for the largest gold deposits in the world, the Witwatersrand deposits of South Africa!

With the resurgence of interest in uranium prospecting in the late sixties, a new deposit type emerged in northern Saskatchewan: the Athabasca-type unconformity uranium deposits. These are now the principle source of uranium in Canada and have enabled the country to become the world's leading uranium producer. The deposits all occur at or just below the base (unconformity) of a major sequence of essentially flat-lying mid-Proterozoic (around 1500 million years old) sandstone beds called the Athabasca Basin and a highly deformed early Proterozoic and Archean granitic and sedimentary 'basement.' This sandstone was deposited in what was probably series of shallow lakes that shrank during dry periods, then reflooded and merged, leaving paleostreams and ancient beaches trapped in the geological record. Some of the early Proterozoic basement rocks include graphitic shale, which played a key role in the deposition of uranium. Uranium was leached from the basement rocks, transported in major faults and shear zones in the basement, and probably along the unconformity surface, and accumulated where the graphitic shale, which contains abundant reductants such as methane, caused it to be come insoluble. The deposits are high grade (90.5% U_3O_8) compared with most other types, but small (0.5-3.5 million tonnes). They are compositionally quite variable. All contain poorly crystallized pitchblende and coffinite, and some, such as the Key Lake deposit, are polymetallic, with recoverable amounts of nickel, cobalt and silver in minerals that are more typical of vein silver deposits than of magmatic nickel deposits. Although most of the deposits discovered in the 1970s are near the margin of the basement, recently there have discoveries made and extensive exploration undertaken below more that 100 metres of sandstone. Canada has several other historically important types of uranium deposits. Cobalt-type arsenide-uranium-silver veins in the Great Bear Lake region produced uranium, silver, copper and cobalt from 1938 until 1955. Pegmatite dykes are the host to past uranium production in the Bancroft, Ontario camp. Volcanic-associated uranium is present in the Kitts Michelin region of southern Labrador, but has not been produced.

Unconformity-type unranium deposits in the Atabasca region of Canada

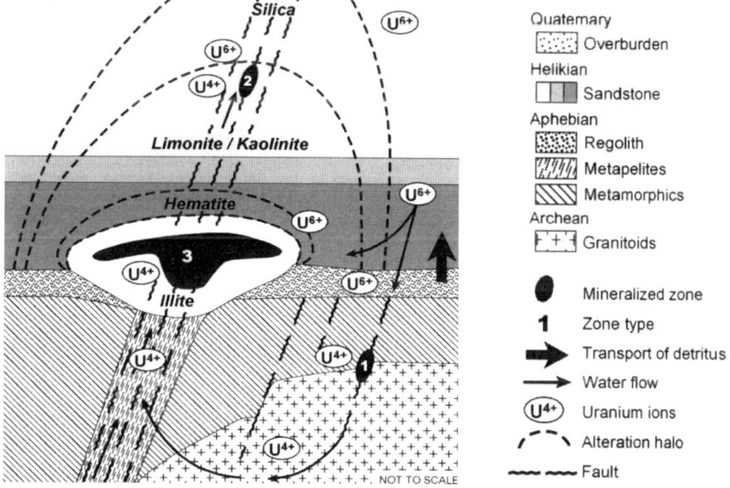

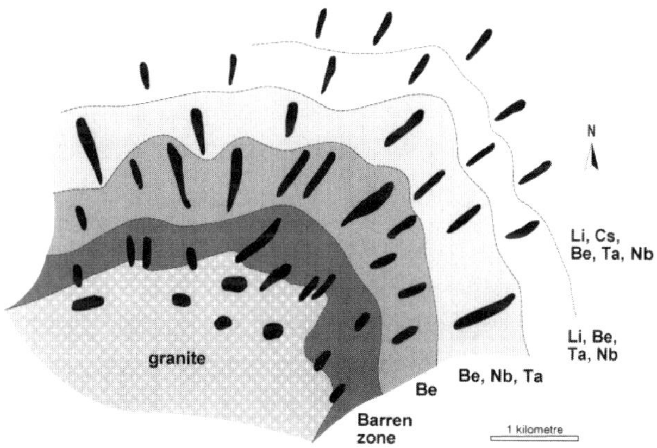

Among some people uranium gets a bad rap due to its use as the explosive material for atomic weapons and yet these folks tend to forget that it has most beneficial uses for mankind, principally as the fuel for nuclear reactors which deliver about 15% of the country's electricity. Canada is currently the largest producer of uranium in the world, although Australia has the larger proportion of the world's known deposits. In 2006 of the seventeen countries that mined the element,

Mine finder Gunnar Berg at the Gunnar Mine, 1934
DAC

Canada produced 28%, followed by Australia with 23%. The term 'yellowcake' was originally given to uranium concentrate, although the colour and texture today can range from anything through dull yellow to almost black.

Early interest in uranium in Canada took a back seat to the work of Gilbert and Charles LaBine who discovered radium at Great Bear Lake in the Northwest Territories in 1930. Gilbert listed his occupation in *Who's Who* as "mine maker." He had attended a lecture in 1913 by Willet Miller, former Ontario Provincial Geologist and the man who recognized the nature of the silver ores at Cobalt. On this occasion he had seen pitchblende and he was one of the few prospectors who could recognize the mineral. His interest at the time was radium and he was not concerned with the uranium oxide found with this mineral. Interest in uranium languished until twelve years later when the scientific community felt that uranium was the key ingredient to the power offered in nuclear fission. The National Research Council of Canada established a research project to use uranium with heavy water as a moderator to produce plutonium in an atomic reactor. Eldorado Mining and Refining Ltd. provided the uranium from its Port Radium Mine and this was followed in 1955 by product from the Gunnar Mine, located on a peninsula on Lake Athabasca, which operated until 1963.

Two areas in Ontario drew a flurry of activity in 1949 as prospectors sought uranium. Bancroft has always been a prime spot for rock hounds. The country around the Hastings County village on the York River is host to 1600 different minerals, some of which are quite rare. The place is due north of Eldorado, site of the first gold mine in Ontario, and Marmora, which had a huge iron deposit. Exploring in Faraday Township, prospector Arthur Shore found what would become the largest property in the vicinity: the Faraday Uranium Mine. A staking rush took place, work was carried out on a hundred sets of claims and this shook down into four uranium mines. The two other significant mines founded hamlets which still exist. Cardiff was set up by Dyno Mines and the Bicroft Mine, the second largest in the camp, had its gathering of houses. A drop in world prices and costly operation in narrow veins saw the end of the Bancroft mines by 1963, although the Faraday reopened for a short time between

1976 and 1982. More than 3 million pounds of uranium oxide were produced in the small camp.

The Bicroft Mine, Bancroft, Ontario, 1956
Denis Newman

The first significant uranium discovery in Ontario took place at what is now Elliot Lake, off the Trans Canada Highway at Blind River between Sudbury and Sault Ste Marie. Sault hotel owner Aimé Breton passed a Geiger counter over ore samples in the Mining Recorder office and the machine came alive with clicks. The sample had come from Long Township on the north shore of Lake Huron. Breton and his partner Karl Gunterman staked some claims in the area and found that there were other claims which had been staked before them. Seeking to sell the ground, they met geologist Franc Jobin who traversed the claims but did not find any uranium with prospects of commercial strength. Later Jobin theorized that the surface ore had been leached away and that drilling was necessary to be sure. By then he was working for American promoter Joe Hirshhorn, always a man to take a flier on a promising venture, who funded the project in 1952.

Pronto Mine, Rio Algom, Elliot Lake, 1958
Denis Newman

The gamble paid off; commercial grade uranium was located between Huronian sedimentary deposits and the basement rock below. The mineralization was contained within a pyritic quartz pebble conglomerate in paleochannels within granitic rocks. Jobin found a Geological Survey of Canada paper published 27 years before that showed this area of contact forming a rough shape of the letter 'Z,' running for thirty miles on each bar. In an operation calculated

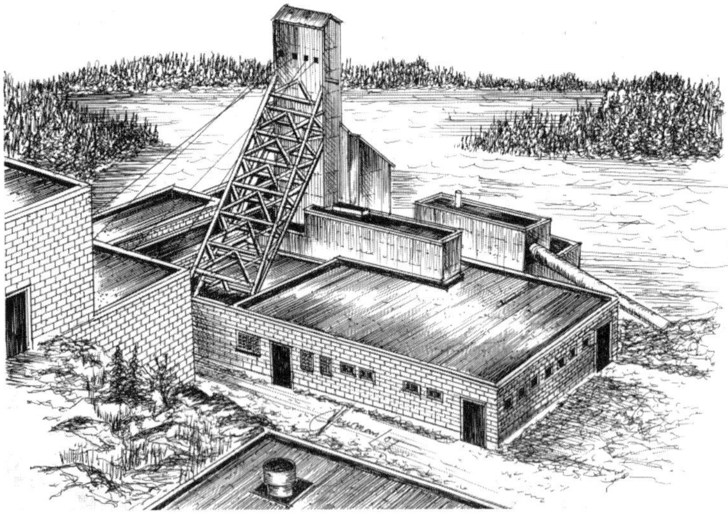

Denison #2 shaft house, Elliot Lake, 1957
Denis Newman

to avoid notice from other mining companies, Hirshhorn flew in seventy-five men who staked wherever the Geiger counter sprang into action. In 1953 the sudden interest in the area was noted and prospectors staked 8,000 claims. One group was led by Arthur Stollery, who staked what became the biggest property in the camp, the Denison Mine, for Stephen Roman. Within five years after Jobin drilled the area there were twelve uranium mines in the Algoma district, eleven of which had their own mills and in total capable of treating 35,000 tonnes a day.

First into production was Hirshhorn's Pronto Mine and activity in the area was so intense that the mines themselves worked on a road to access the main highway without waiting for the province to wake up and get in on the bonanza. By 1954 the Ontario government was co-operating with the mining companies to build a model town, Elliot Lake. Five years later the new municipality had thirty thousand residents and called itself "The Uranium Capital of the World." The twelve mines had a total of 10,000 employees. Hirshhorn eventually sold out to Rio Algom, now part of Rio Tinto. The mines in its stable were Stanleigh, Nordic, Lacnor, Quirke, Panel and Pronto, which in total mined more than 100 million pounds of uranium. Roman's group included the Denison, Stanrock and Canmet mines. The other independents were the Milliken, Buckles and Spanish American. Total production from all the mines was more than 270 million pounds.

Somebody once said that anyone who was in Elliot Lake from the beginning must have a liking for roller coaster rides. In 1959 the model town experienced the unwelcome thrill of the downward phase. The American Atomic Energy Commission felt it had enough uranium stockpiled and unilaterally cancelled its contracts. Hundreds of miners left the area to work in places like Wawa and Sudbury and many homes were boarded up. By 1960 five Elliot Lake mines had closed, and the federal government was forced to prop up the local economy with make-work projects and pay to have uranium stockpiled by mines still operating. Eventually the market picked up, the town was revitalised, and contracts came on stream from the United Kingdom and others involved in atomic energy for peaceful, electric generation purposes.

With other regions able to mine uranium more cheaply and the inevitable near-exhaustion of the ore bodies, most area mines closed by 1990 and the last, the Stanleigh Mine, was finished in 1996. Over their lifetime the mines averaged a respectable two pounds of uranium per ton. The Denison interests reinvented themselves by offering mine closure services and once again it seemed that Elliot Lake was on the way to becoming a ghost town. But using aggressive promotion, the town, which enjoys fine modern homes and a beautiful setting, became a retirement haven and thrives today. Uranium prices are now high enough that the ore at Elliot Lake is considered, by some miners, worth a second look.

Saskatchewan stood next to Ontario in the hunt for uranium. No one in the early fifties realised that it would become the dominant province for the ore in Canada and maintain that position. The Saskatchewan government organized its own Mining Development Corporation to promote and develop promising ore deposits. Gulf Minerals began to construct its mine at Rabbit Lake which lay in altered basement rocks at the edge of sandstone. In addition, there were three other major areas of high interest: Cluff Lake, which is now mined out, near the Alberta border; Midwest Lake in the northeast; and Key Lake, to the south and midway between the Alberta and Manitoba borders. All prime uranium-producing areas were situated within and adjacent to the Athabasca Basin. By December 1968 fifty companies were involved in a huge staking rush. For the first time in Canadian history oil companies joined the hunt for uranium in a big way. By 1975 the Rabbit Lake Mine had begun operation, the start of what would become the major uranium mining location in the country.

The New Democrats of Saskatchewan sought to phase out uranium mining in the early nineties, but subsequent governments reversed this nearsighted decision. Today the two principle mines, McClean Lake and McArthur River, possess ISO 14001 environmental certification.

Gunnar Mine, Lake Athabasca
Sask. Industry

Eldorado Nuclear Mine, Uranium City
Sask. Industry

Cigar Lake has one of the largest uranium deposits in the world.
Sask. Industry

McArthur River Mines, Cameco
Sask. Industry

The McClean Lake Mine dates from 1999 and production is being ramped up to a possible 3,640 tonnes of uranium per year and will go from four open pits to underground mining. The McArthur River Mine has very high grades at 23% and could operate at 10,000 tonnes per year. The Cigar Lake Mine was due to start up in 2007, but severe flooding set the project back until possibly 2010 with what appears will be a final start-up tally of $1 billion. Since the $508 million project has proven reserves of 226 million pounds of uranium and an estimated mine life of 15 years, the intent to have the property in operation by 2010 has been welcome news for world demand, which is outstripping current production. Other projects to come on stream are the Midwest and Dawn Lake mines.

There is much interest in uranium today due to high world demand. The eastern area of the Central Mineral Belt of Labrador has seen much attention, especially in the Michelin deposit, which is estimated to hold 18.3 million pounds of the mineral. Exploration is ongoing in the western and southern parts of Newfoundland. Manitoba is seeing renewed activity in the northwest around the Saskatchewan border, an area that was previously of interest in the seventies. Neighbouring Saskatchewan remains the main focus of uranium exploration in Canada. At present, production is around 30 million pounds annually from three locations. The Athabasca Basin is the best bet for new discoveries; a recent tally showed thirty-five companies actively exploring in the Basin. Two thirds of all mineral exploration rights in Alberta applied to uranium. They are mostly in the northeast unconformity and vein-type deposits and in the sediment-hosted areas of the central and southwestern portions of the province. There is some interest in the Wernicke Mountains of northeastern Yukon and in the Thelon Basin of the Northwest Territories, where radioactive sandstone boulders are found. The Pyramid showing, 300 kilometres southeast of Yellowknife, has been sampled for uranium on and off since 1934 without any significant discovery.

One of the reasons the hunt for uranium has intensified is that for several years the nuclear industry has been living off inventories and converted weapons stock, which are now very close to being used up. In Saskatchewan the main thrust is on the search for unconformity-type deposits in the traditional pay off area, the Athabasca Basin. Geophysical surveys and interpretive software are now standard exploration tools. Geologists are aware that uranium mineralization can occur up to 300 metres away from the conductor. Favourable targets are expected to be found much deeper than in earlier successes. Much exploration emphasis now is on integrating uranium deposit exploration geology with geochemical methods. Geochronology (age dating) and isotope geochemistry aid in identification of regions likely to host uranium deposits.

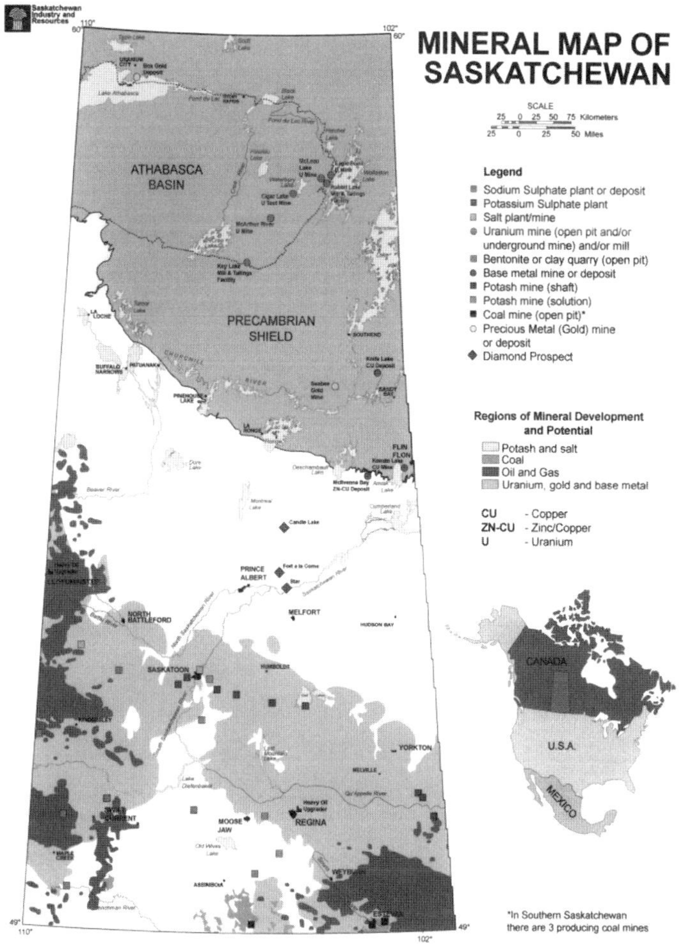

In common with other resources, uranium has been a victim of economic cycles, but as an energy source it is definitely on the upswing. Environmentalists are coming onside with a commitment to nuclear power because it does not contribute to air pollution or greenhouse gases. The market is growing for fuels that do not contribute to global warming. At last count worldwide there were 435 reactors operating and as many as 92 either in construction or planning stages — good news to uranium miners. Historical sources of energy such as oil and coal have seen substantial hikes in extraction costs, and as a result, the large capital investment for new nuclear energy installations have become more reasonable to power producers. Uranium has become one of the biggest stars of the commodity boom.

Black Rock

Your Basic Intrusions

Intrusions, bodies of melt that originated in the upper part of the earth's mantle or deep in its crust, contain a wide variety of ore deposits; all tend to be very large, compared with stratiform deposits. They occur in all ages of rocks, but in general, the largest ore deposits are in Proterozoic or younger intrusions. They may be divided into two broad classes, each with its unique set of deposit types. Those that formed in rocks rich in iron, magnesium and calcium silicate minerals (very dark in colour) are termed 'ultramafic' (most magnesium-rich) and 'mafic' intrusions with formal names such as — in order of decreasing amounts of magnesium — dunite, peridotite, pyroxenite gabbro and anorthosite. All of the melts that formed these rocks originated in the upper few hundred kilometres of the mantle, and most formed in zones of continental break-up and sea floor spreading. Depending on the composition of the original melt, the entire spectrum of these rocks may be present in one suite. With time, dunite, pyroxenite and peridotite will first precipitate from a primitive, mantle derived melt. These rocks may be followed by gabbro and anorthosite.

The second group of rocks, rich in silica, sodium potassium and aluminum are termed 'felsic' and include terms such as granodiorite, granite and syenite. These formed typically from the melting of a down-going slab at a zone of collision, apparently between oceanic and continental plates, such as is occurring along the west coast of both North and South America.

All of the deposits that form in mafic and ultramafic rocks are termed 'mafic-magmatic' deposits, and all share many aspects of a common genetic model. There are four groups of deposits worth examining in the Canadian context. By far the largest and most economically important are the magmatic nickel-copper-platinum group element (PGE) of deposits. These, along with vein gold and VMS deposits, define the core of the Canadian mining industry. Although more restricted in size and economic contribution, the other deposits associated with this group of rocks include the asbestos deposits (and related talc deposits) of the Eastern Townships of Quebec, Northern Ontario and British Columbia, the iron-titanium deposits of the Lac Allard district of central Quebec, and a few chrome producers in B.C. and Ontario.

Nickel-copper PGE deposits occur in three super-giant districts in Canada: Sudbury, in a virtual class by itself; Thompson, Manitoba; the Cape Smith belt of northern Quebec; and the Voisey's Bay district in Labrador. There are numerous smaller deposits scattered through the Shield. The processes by which these deposits formed share many common aspects. First, all mafic, and particularly ultramafic, rocks contain higher than average 'background' contents of nickel, copper and the PGEs. Ultramafic rocks, for example, contain background nickel content of about 2000 parts per million (ppm), and mafic rocks contain about 150 ppm. These elements normally end up being sequestered in various silicate or oxide minerals; for example, nickel goes into olivine, an Mg-Fe silicate, one of the first minerals to crystallize in a cooling melt. In order to make a nickel-rich ore deposit, it is critical that the melt, before any silicate minerals have been crystallized, becomes saturated in sulphur. This usually happens only if the melt encounters and engulfs a sulphur-rich rock (such as iron formation) when it is rising from the mantle into the crust. The sulphur combines with iron in the melt and incorporates much of its nickel along with the copper and PGEs into a sulphide-rich liquid that separates quickly from the parent silicate melt, much like oil in water. The 'oil' in this case is formed by tiny sulphide droplets dispersed in the 'water,' which is the silicate melt. All of this happens at very high temperatures, above 1200 C. These droplets have a very high specific gravity relative to the silicate liquid, and they quickly settle to the bottom of the magma chamber, accumulate as pools of liquid sulphide in troughs and embayments at its base and eventually solidify to form an ore deposit. Some droplets remain frozen in the crystallized melt; others accumulate to form massive sulphide. If the original melt had less nickel, or the nickel had already been 'stolen' by olivine, then the sulphide deposit will be copper and PGE rich. Too much sulphur added to the melt makes a lower grade of the deposit, and if only a little is added, then the resulting sulphide droplets are tiny but exceptionally rich in PGEs.

Although virtually all of Canada's magmatic nickel-copper-PGE deposits form by some variation on the theme just presented, there are some interesting circumstances that seem to have enabled some districts to become exceptionally large. The famous Sudbury basin is one such interesting case. The 'basin' is an elliptical ring formed by a mafic intrusive complex injected into a complex set of basement rocks (including both an Archean greenstone belt and the Huronian rocks that elsewhere host the Elliot Lake deposits). The basin is filled with sedimentary and volcanic rocks. All of the deposits sit at the base of this intrusive complex, and their presence seems to fit with the model just described. In the 1960s a group of planetologists (geologists interested in extraterrestrial bodies) began to find compelling evidence of a giant meteorite feature at Sudbury: glassy melted country rocks, impact shatter cones, and a thick unit of fall-back material called the Onaping Formation. It is now generally accepted that Sudbury was the site of impact of a large asteroid that, when it hit, compressed the crust and upper mantle. As the latter rebounded, depressurization caused the mantle to melt and mafic melts were injected quickly along the ring fractures caused by the impact. Although the process of formation of the nickel-copper PGE sulphides is, in detail, the same as for other deposits, the huge amount of energy that started the whole process came from a meteorite impact!

Canada's other major nickel districts all seem to have formed at the margins of giant rift structures that may, in some cases, have preceded the formation of major ocean basins. The Thompson belt and the Cape Smith deposits, although separated by nearly 2000 kilometres, actually formed at exactly the same time, and through tracing these sequences (with the help of geological surveys of Hudson's Bay), they can be shown to be parts of the same rift feature. The recently discovered deposits at Voisey's Bay share many of the same features as the Thompson and Cape Smith deposits but formed about 600 million years later. The North American Palladium deposit at Lac des Isles, north of Thunder Bay, Ontario, is another special case of the same theme. It contains only a tiny amount of sulphide, but these tiny grains are extraordinarily enriched in palladium, making them economically recoverable.

The asbestos deposits of the Thetford Mines area in the Eastern Townships of Quebec are formed from an ultramafic rock that, just after its emplacement, was subjected to reaction with a huge amount of water. All of the olivine was transformed into hydrous minerals such as crystolite, one of the asbestiform minerals. The iron-titanium deposits in the Lac Allard, Magpie Mountain and Newboro Lake areas of Quebec all occur in Proterozoic mafic intrusions that are at the most 'evolved' end of the mafic spectrum, anorthosite. Here, all of the magnesium and most of the calcium and sodium have already formed silicate minerals, leaving behind a liquid exceptionally rich in iron and titanium. The same immiscibility process takes place, but iron-titanium oxide minerals (principally the mineral ilmenite) accumulate to form a deposit. A related process in less primitive rocks forms chromite zones; these deposits are relatively rare in Canada, but have been mined in British Columbia, Quebec and Ontario.

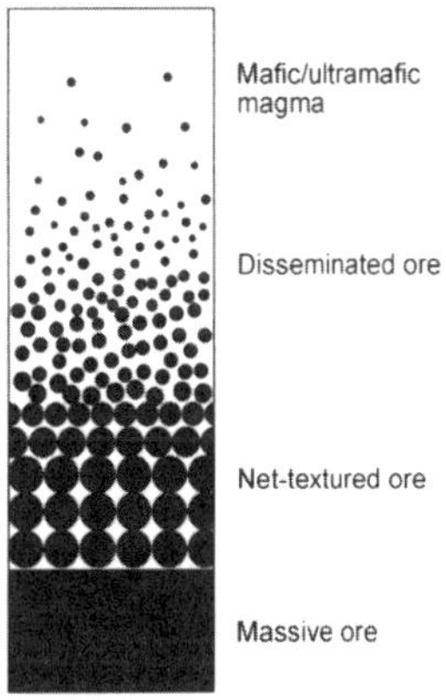

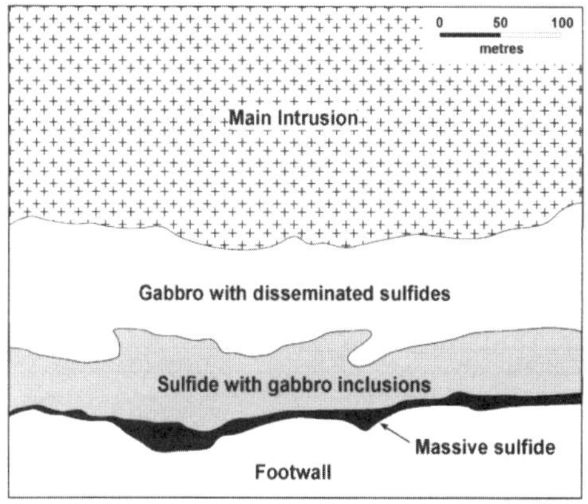

One thinks of intruders as folks who come unwanted or uninvited, forcing themselves abruptly on others. In geologic terms intrusive rocks are not necessarily unwelcome as they can be the source of sought-after mineral deposits. Intrusive rocks start off as molten rock or magma which works its way toward the surface of the earth and solidifies before it reaches the surface. Often called plutons after Pluto, the god of the underworld, these igneous rocks find a place in the overlying or country rock, either forcibly or gently. As the song says: "Something's gotta give." Forcible intrusions move in and fracture the country rock, allowing mineral-bearing hydrothermal fluids to circulate. Providentially waterborne minerals can be deposited in these fractures in the rock. Intrusive rocks come in many geometrical forms, providing a fair share of puzzles to be unraveled by exploration geologists.

The Raglan Mine sits way up north in the Nunavik territory of Quebec, accessed by road from Donaldson, and ships its product from the

Kenneco exploration crew in Kiglaplait, Labrador, 1970: David Barr, David McAuslan and the pilot
D. Barr

concentrate storage at Deception Bay. Present owner Xstrata inherited a deposit of massive sulphide lenses stretching over 55 kilometres, which began producing in 1997. The complex is supported by three underground mines and an open pit, and is expected to have a thirty-year life. The fly-in operation processes 2,800 tonnes per day and has highly profitable by-products of palladium, platinum and cobalt as a bonus.

Northern Labrador is one of the last great frontiers for explorationists in Canada. The Voisey's Bay area near the tiny settlement of Nain, at the northern tip of Labrador, hard by the Quebec border, was mapped and sampled by provincial geologists in 1985. But it was not until nine years later that a couple of junior company prospectors searching for diamonds noted visible gossans, orange-coloured iron-rich crusts of decayed rock, and staked the ground on a hunch. Subsequent drilling showed pentlandite, similar to that found in the Sudbury basin, and brassy crystals of chalcopyrite, copper ore. Over the next two years it was determined that the property was a world class magmatic nickel sulphide deposit that consisted of an ovoid chamber and the Eastern Deeps chamber. The prize was sought by several major mining companies and eventually secured by Inco. On further exploration the deposit was determined to comprise 30.4 million tonnes containing 2.85% nickel, 1.65% copper and 0.142% cobalt. The announcement of the discovery in 1994 set off a huge exploration rush, with expenditures over the next two years more than eight times what had been spent in the province the year before. The giant deposit was found by surface discovery, and the aftermath was protracted wrangling between the province, anxious to protect and maximize its resource benefits, and Inco, which wanted to develop its prize as expeditiously as possible. In 2006 further exploration two kilometres west of the main

Raglan concentrates awaiting shipment to Sudbury
Xstrata Nickel

The concentrator complex, Voisey's Bay
VBNC

deposit added significantly to the reserves. The plan is still to partially process the finished nickel in Newfoundland.

Campbell Resources has a small mine that opened in 2005 in McKenzie Township, seven kilometres from Chibougamau, Quebec, which has its mineralization in sulphide lenses within an intrusive related base. The adjacent Copper Rand mine has its own mill and also produces copper.

Move west to Ontario and to the great nickel camp of Sudbury, the second largest producer of nickel in the world after the Noril'sk complex of Siberia. Located west of North Bay in northeastern Ontario, nickel-bearing rocks were found in 1853 as the Canadian Pacific Railway was being built across the country. No one paid any serious attention to the deposit until the early 1880s. Since that time there has been much debate as to the origin of the vast nickel copper deposits located in the Sudbury basin. There are two theories as to the cause of an event that has been placed at 1.85 billion years ago. One is that a huge volcano exploded and resulted in the 60 by 25 kilometre broadly elliptical scar, ten kilometres deep, which is the basin. The other more commonly accepted view is that an asteroid slammed into the earth and the fracturing from the impact left many channel ways for the ore-bearing fluids from below. Whatever the origin, the nickel was not easy to extract from the rock and International Nickel did not start mining until 1902 and Falconbridge commenced

Mines and main deposits in Québec

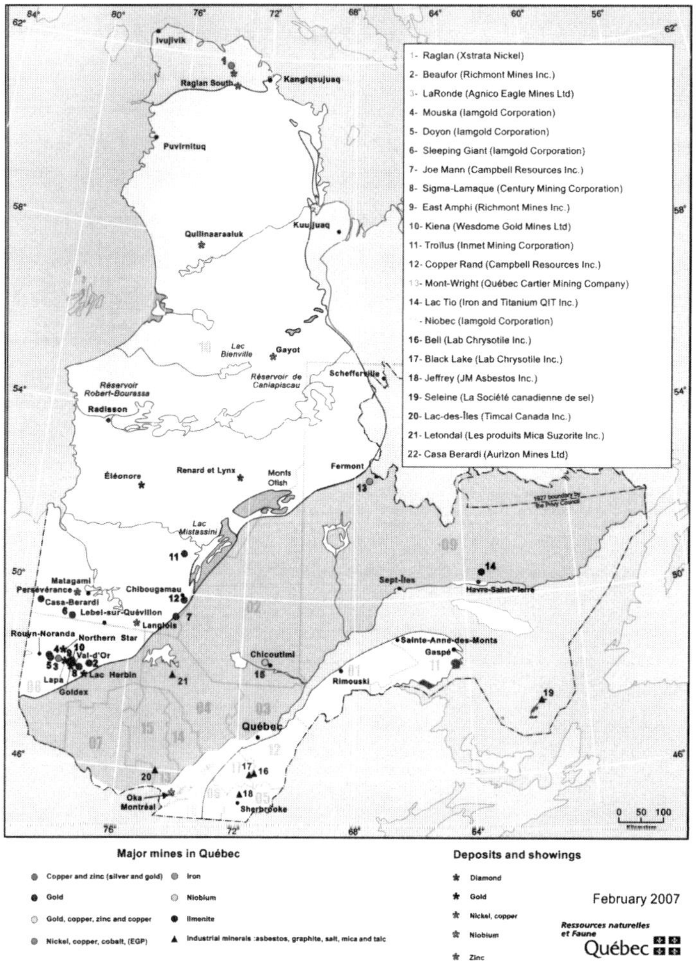

Creighton # 3 Mine, CVRD, Sudbury
Denis Newman

operating in 1928. Mines have come and gone since that time, and at present 19 mines operate around the rim of the basin with more than 50 million tonnes of rock being processed each day. Nickel production from the Sudbury Basin has been a constant throughout the twentieth century.

Deposits within the basin continue to be found. A recent one, discovered in 2001 by Falconbridge geologists, was the Nickel South deposit on the east of the basin, close to the Sudbury airport. Falconbridge, now Xstrata, has other growth projects to the west of the basin rim such as the Fraser Morgan, due to commence production in 2009, and the Onaping Depth. Inco-CVRD has four more mines in development, all east or south of the basin. These are the Victor and Blezard, with the Kelly Lake and Totten projects in the feasibility stage. CVRD started producing from the Garson Ramp in early 2007. To these veteran miners, add First Nickel's Bower deposit; (shaft sinking and now in production) FNX Mining's Podolsky copper-PGM deposit; new discoveries under the old Levack Mine workings; and development at Ursa Major's Shakespeare nickel deposit. When brought into production, the Shakespeare property will have a handsome reserve of 11.26 million tonnes grading 0.33% nickel, 0.35% copper, 0.02 % cobalt, 0.333 grams per tonne palladium and 0.9 grams per tonne precious metals. There is no dearth of activity in the Sudbury mining powerhouse. The Timmins

Frood Mine # 3 head frame, Sudbury
Denis Newman

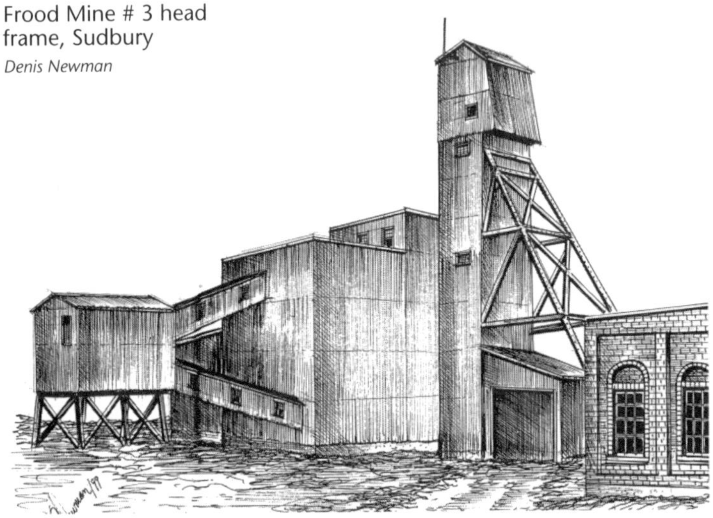

camp now has the Redstone Mine of Liberty Mines, which has been producing since mid 2006. This is in addition to Xstrata's Montcalm copper-nickel deposit which commenced production in 2004, shipping ore to the Kidd Creek metallurgial complex at Hoyle near Timmins.

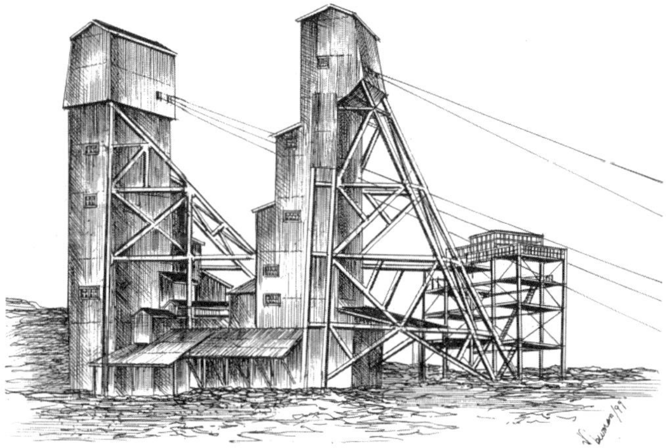

Stobie Mine # 7 and head frames, Sudbury *Denis Newman*

North American Palladium often refers to itself in print as NAP but there is nothing sleepy about the mining complex and camp accommodation at Lac des Isles, eighty-five kilometres northwest of Thunder Bay, Ontario. The property deposit is hosted in mafic-ultramafic intrusive. It holds the only primary palladium producer in Canada and one of the largest deposits of the precious metal in the world. The plant until recently has been open pit but has gone underground to access higher grades as feed for the 15,000 tonnes-per-day mill. The output is truly polymetallic: in addition to a palladium grade of 2.33 grams per tonne, which delivered 25,128 ounces of palladium in 2004, there were also by-products of 25,679 ounces of gold, 7,836,183 pounds of copper and 4,320,970 pounds of nickel.

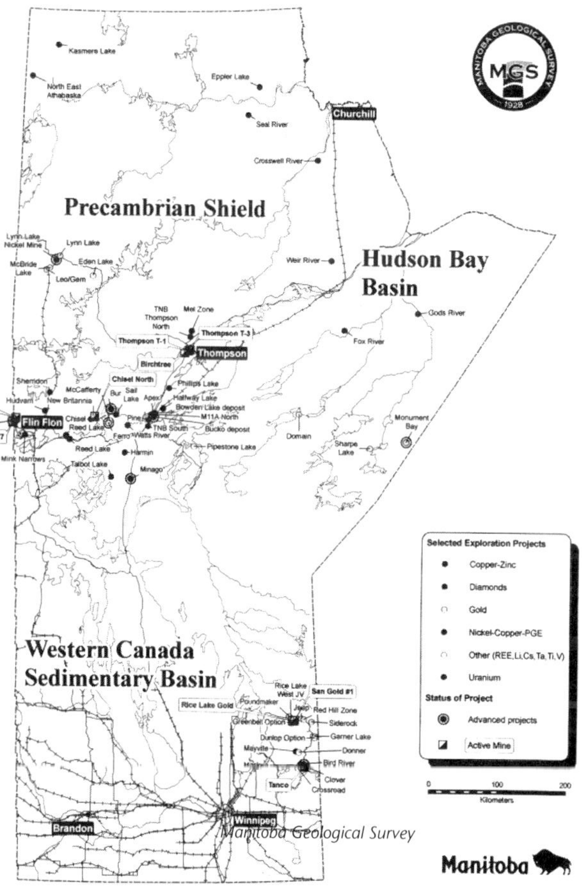

Manitoba Geological Survey

Manitoba had base metals added to its mining portfolio with Hudson Bay Mining and Smelting at Flin Flon in 1927, and the Sherritt Gordon Lynn Lake property, which started up in 1931. Actually, initial penetration of the area was by famed geologist and

Dr. John Thompson outside a prospector's cabin in 1956, in the location of present day Thompson, Manitoba
PDAC

explorer Joe Tyrrell back in 1896. In both cases start up costs and the capital required for infrastructure in undeveloped land were enormous. This was also the case for Inco right after World War II. In 1946 company geologists were looking for extra feed to supplement the output in Sudbury. Using the newly emerging electromagnetic surveys and what would eventually amount to 720 kilometres of core drilling, the determined nickel miner searched for six years before a promising deposit was located. It was ten years after the start of explorations before a decision to build a mine at what became Thomson, Manitoba, named after a longtime company chairman. The geology lay in the roots of an ancient mountain system, and the ore was lower in copper content than that found in Sudbury. Five years passed from the construction decision to the 1961 start up, as the new town location, 700 kilometres northwest of Winnipeg, required construction of a power plant seventy-two kilometres away on the Nelson River and a forty-two-kilometre branch rail line to the Hudson Bay Railway. During that period several other major deposits were located. The area is still going strong with 120 million pounds of nickel produced at Thomson in 2006 alone.

Hidden in the Rock

Porphyries

Mineral deposits in the 'granitic' suite of intrusions have different mechanisms of formation, resulting in a completely different group of commodities recovered from them. Canada is modestly endowed with supergiant deposits within porphyries, but many occur in other parts of the world, most notably the 'ring of fire' that fringes the Pacific Ocean. One class of ore deposits dominates over all others in this group, the porphyry copper, copper-gold, copper-molybdenum, molybdenum and tungsten-molybdenum deposits, grouped here as the 'porphyries.' By far the largest amount of the world's copper comes from these, along with a substantial amount of the world's gold, and virtually all the world's molybdenum. The porphyry types of deposits share many common elements; they are all related to major (kilometres wide) intrusions and associated volcanic rocks, and they were all generated by a fluid that emanated out of the melts that formed these intrusions and volcanoes. The deposits all consist of strongly altered rocks that contain sulfide disseminations (hence the name 'porphyry,' which relates to the distribution of the sulfide minerals into discreet grains and millimetre- to centimetre-scale blebs) and veins of base metals and iron sulfide minerals.

Regardless of metal combinations, these deposits are truly giants. The deposits range in size from several hundred million to over 17 billion tonnes; the size of a deposit is dependent on the lowest grade acceptable for mining (the cut-off grade) and this is dependent on metal process. Sizes may expand or contract depending on current metal prices. The copper group typically contains from 0.3 to 1% copper, the molybdenum group from 0.1 to 0.5% molybdenum, and the gold-rich systems contain from 0.15 to 1.5 grams per tonne gold. The deposits of South America, Indonesia, Asia, the southern U.S. and Nuguini, the Philippines, have undergone a process called supergene weathering, which has distributed the copper into high-grade zones. These are treated by the heap leach process (solvent extraction and electro-winning) where the broken ore is placed on large leach pads, sprayed with water, and the copper leached out and recovered from the water as it drains out of the rock pile. Unfortunately, Canadian deposits are not as amenable to this treatment, as the intensive sub-equatorial weathering process has not affected them.

The majority of the economic deposits in Canada occur in the Canadian Cordillera, in rocks of two age ranges, 170 to 220 million years (mid-Triassic to mid-Jurassic) and 36-85 million years (Eocene to Late Cretaceous). Those in the older group are copper-gold deposits; those in the younger group are copper-molybdenum deposits. Only a few economic deposits are known outside of the Cordillera. The most notable is the Paleozoic (Carboniferous) Gaspé Copper deposit in Quebec, the Tribag copper-molybdenum deposit in Ontario (mid-Proterozoic) and the McIntyre Mine at Timmins (Archean).

Although all are related to intrusions, many have formed well above them, in volcanic and sedimentary strata, and the depth to an intrusion may not be known. However, most, such as the deposits at Gibraltar, Granisle, Brenda and Mount Polley of B.C., occur at least in part in an intrusive suite. Virtually all of the deposits are associated with breccias, which resulted from intense, forceful fracturing of the rock due to extreme fluid pressures. The vein relationships are complex, and original features of the rock largely obscured by intense reaction and alteration with the ore-forming field. The exact source of fluids for porphyry deposits is enigmatic, but it is clear that these fluids have interacted with a major intrusion.

A less common but important deposit type that is closely related to the porphyry group is the tin-tungsten, lead-zinc-copper, copper and iron-gold scarn group of deposits. Almost all of these are formed in the Canadian cordillera. Formed from hot fluids that escaped from a felsic intrusion, these fluids reacted with carbonate-bearing rocks such as limestone, which not only caused the metals to precipitate, but also formed, resulting in very unusual mineral assemblages. The metal assemblage is dependent on the type of fluid that emanated from the intrusion, as well as the distance away from the intrusion where the reaction occurred. Fluids that interacted a long distance from the intrusion may have cooled and formed lead-zinc-silver deposits, such as the Bluebell Mine in B.C. and the Mount Polley porphyry copper deposit, which also contain some skarn mineralization. Almost 100% of Canadian tin production came from skarn deposits such as Canada Tungsten Mine in the N.W.T. Finally there are the Tasu, Iron Hill and Texada iron mines on Texada Island, B.C. Similar small deposits have been mined at Marmora, Ontario and Bristol, Quebec.

Pegmatic deposits are a distant cousin of the porphyry type, and have been an important source of industrial minerals in Canada. These are sources of lithium, tantalum, niobium, beryllium, tin and tungsten, as well as uranium. There are, to some extent, the felsic equivalents of the mafic magmatic deposits if the mineralization associated with them occurs within the intrusion and is not transported into a fluid phase that may migrate out of the intrusion.

Pegmatite dykes usually form at the margins of their parent intrusions and are exceptionally coarse-grained. The melts that form pegmatite are very rich in silica, potassium and sodium and contain a large amount of trapped water. The water lowers the crystallization temperature of the melt and also accumulates all of the elements from the parent melt (usually granite) that precipitates late in their crystallization sequence. These melts crystallize very slowly, allowing the individual minerals to become very large. Crystals are from the tens of centimetres to metres in width and, in rare circumstances, even up to tens of metres across. Besides being great rocks for collectors and providing spectacular exposures, they provide a unique set of 'high technology' materials that are in great demand by the electronics, pharmaceutical and space industries. Currently the Tanco Mine in south-eastern Manitoba produces many of these rare metals and industrial commodities. The Madawaska Mine near Bancroft, Ontario, produced uranium, and the Presissac-Lacorne deposit near Val D'Or, Quebec, produced lithium.

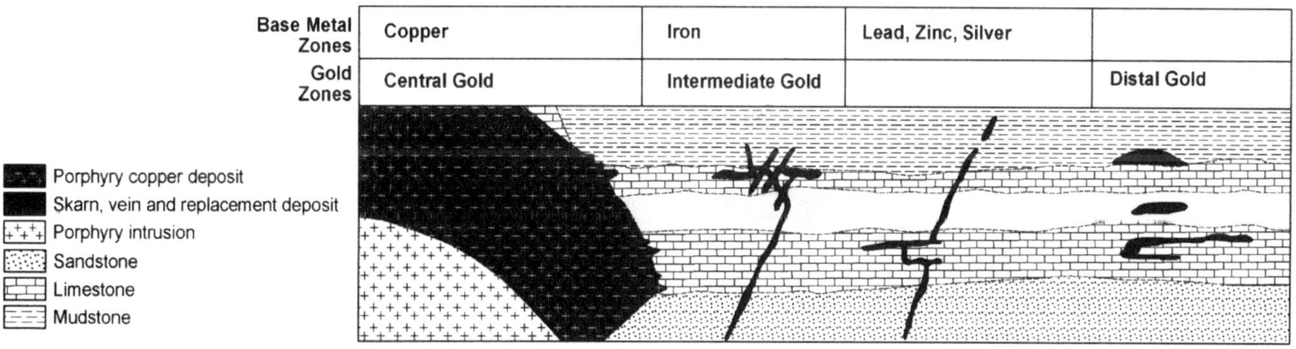

Those who seek minerals in porphyries would be advised to follow the old adage, "Go west young geologist," as this form of igneous activity is found in young rock with large crystals. Deposits are usually large but the trade-off is in low-grade mineralization. The name porphyry comes from the Latin for its colour purple and has associations with royal or imperial qualities dating back to the Romans. In Canada, British Columbia enjoys the lion's share of this rock, which contains the largest resources of copper, significant molybdenum and 50% of the gold in the province.

British Columbia is copper-rich, and mining of the metal commenced in the late nineteenth century. Many mines have been worked in the province over the past 125 years, and there are currently still some porphyry deposits of interest. But it was not until 1962, when Bethlehem Copper was bankrolled by Japanese industrialists, that the province saw its first profitable low-grade copper-molybdenum mine up and running. Twenty-four years later Bethlehem and Lornex mines were combined to form Highland Valley Copper.

Kemess Mine, northern B.C.
Northgate Minerals

Kemess South is a fly-in operation in the delightfully named Toodoggone area in northern British Columbia. The mine is a copper-gold open pit with a 52,000 tonnes-per-day mill. The deposit was located in the early sixties and is now owned by Northgate Minerals. The concentrate is trucked 380 kilometres to a rail spur at Mackenzie and then shipped all the way to the Horne smelter at Noranda, Quebec. So far the mine has processed 145.0 million tonnes grading .20% copper and some gold and silver as by-products. The south pit will likely be mined out by 2008; the north pit will still provide feed for some years to come.

In west central B.C., 123 kilometres southwest of Houston, the Huckelberry copper mine is owned by Imperial Metals and three Japanese smelting companies. It went into production in 1997 after having been discovered thirty-five years earlier by Kenneco. Extensive stream sampling gave the best clue to copper mineralization on the site. The feed

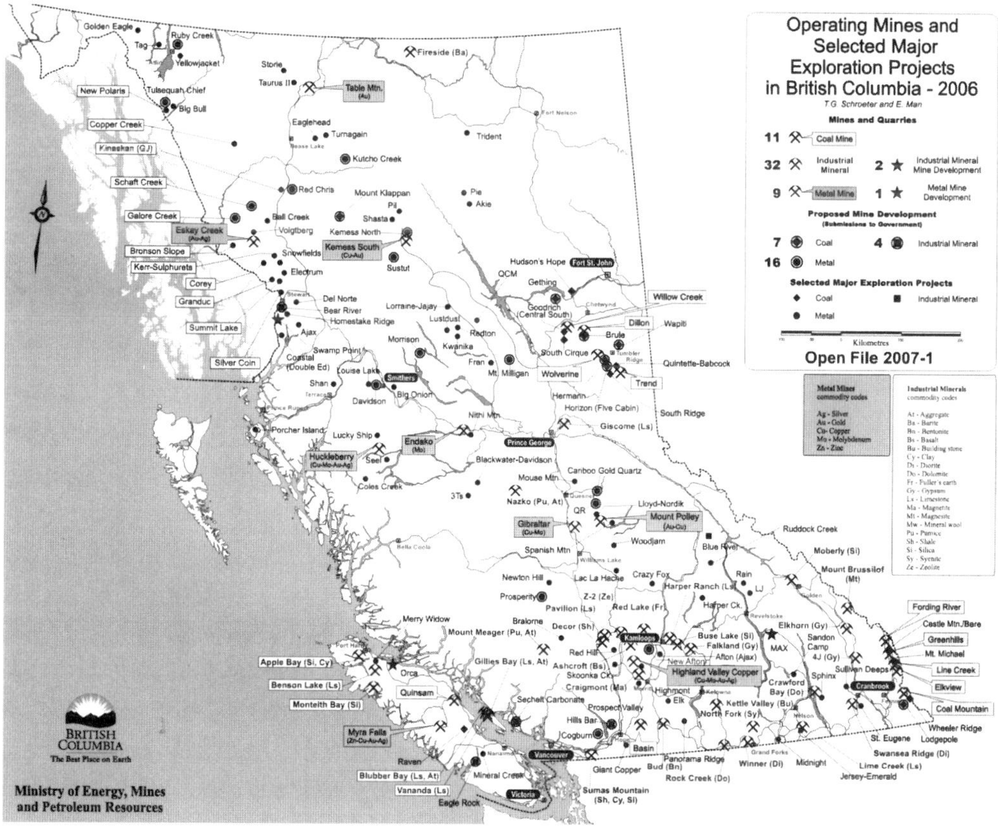

from the open pit is put through the mill at an average 19,500 tonnes per day. By the end of 2006 36.0 million tonnes had been mined at 0.95% copper and .01% molybdenum.

South of Huckleberry, Gibraltar Mine is in the Cariboo ranching region near Williams Lake. The current mine opened in 1972. The property closed in 1992 due to low metals prices and was acquired by Taseko Mines which reopened the open pit and plant in 2004. Gibraltar has mined 3.5 million tonnes of .32 % copper and .81% molybdenum to the close of 2006. Current mine life is projected until at least 2016.

The deposit of Island Copper, a Vancouver Island property of BHP Billiton, was discovered in the late sixties by prospectors following aeromagnetic survey results from work done by the B.C. Department of Mines and the Geological Survey of Canada. From 1971 to 1995, the mines produced 366.7 million tonnes grading 0.37% copper, 0.011%

The Huckleberry Mine is in a beautiful location.
Imperial Metals

molybdenum as well as 0.1 grams of gold per tonne and 0.80 grams of silver per tonne. The mill treated 50,000 tonnes per day and was a world pioneer in waste disposal.

The mine used 'engineered' submarine tailings disposal. Nearly 360 million tonnes of fine ground tailings were placed on the sea floor without adverse effects on benthic life.

Imperial Metals held a big celebration when the Mount Polley open pit mine was reopened in 2005. Originally constructed in 1997, the mine located in south central British Columbia northeast of Williams Lake, had closed four years later due to low metals prices. Production to 2006 was 50.0 million tonnes of 0.22% copper. Scheduled for start up in 2008, the Red Chris copper gold project has involved $228 million in costs but should produce 110 million pounds of copper and 75,000 ounces of gold in its first five years. Imperial is looking to acquire current Red Chris owner bcMetals.

The jewel in the B.C. porphyry copper crown is the Highland Valley camp which includes Valley Copper, Lornex, Bethlehem and the original Highland Bell. The combined operation is huge although very low grade and the Teck Cominco property which gives its name to the camp is the largest metal mine in the province. One of the largest open pit copper complexes in the world, Valley Copper is situated 75 kilometres west of Kamloops. From 1995 to 2005 Valley Copper produced 350 million pounds of copper per year plus between 3 and 10 million pounds of molybdenum per year. The mill has exceeded 160,000 tonnes per day although the average is 139,000 tonnes. Currently Highland Valley Copper is the fifth largest producer of copper concentrate in the world. Teck Cominco is spending $300 million to extend the life of the mine until at least 2019. Output from all producing mines at the Highland Valley camp since 1949, when Highland Bell opened, has totaled 1,250 million tonnes of ore grading 0.41% copper and .007 % molybdenum.

Molybdenum can also be the principle metal within porphyry. Geologists usually find the element in copper porphyry, the source of 60% of the deposits and obtain it as a by-product or occasionally by itself in the host rock. The soft metallic silvery white mineral is often found in foliated masses like mica. Moly is used in construction steel, stainless steel and in the so-called super alloys but its use is now in demand in the construction of fuel cell vehicles. Few people realize that the metal is widespread and needed to sustain plant and animal life. The problem is that few deposits warrant the extensive mining, milling or processing necessary for economic recovery. Canada produced 19.3 million pounds of molybdenum in 2005. The world total was 384 million pounds; the principle world miners are the USA and Chile with their huge long-lived porphyry deposits.

Helicopter assisted exploration north of Smithers, B.C.
Imperial Metals

Endako Mine is the star in molybdenum mining in British Columbia. Its location is 190 kilometres west of Prince George. The ground was located by prospectors in 1927, but no work was done until 1965. The mineralization is typically porphyry hosted. The open pits were closed in 1982 due to low prices but reopened in 1986. Subsidiary of Blue Pearl Mines, Thompson Creek Mining has 60% of the property and a Japanese firm, Sojitz Moly Resources, has the remainder. The mine has reserves to 2013 and currently produces a 0.06% grade from its 30,000 tonnes per day mill with a recovery of 78% moly-sulfide to convert to a moly-oxide.

The Brenda Mine, 225 kilometres north east of Vancouver, was controlled by Noranda for its complete producing life. Found by a prospecting family in the 1930s, the deposit waited many years to be explored due to a lack of demand for molybdenum. Named for a surveyor's daughter and a nearby lake, the Brenda commenced operation in 1970 and, in its two-decade production from a 33,000 tonnes-per-day mill, gave up 161.7 million tonnes grading 0.17% copper, 0.04% molybdenum, 0.1 grams per tonne gold and 1.06 grams per tonne silver.

Work break underground at the Max Mine
Roca Mines

High commodity prices have spurred molybdenum production and development plans in B.C. Roca Mines is building its Max Mine with a

Neat core shacks at Ruby Creek, B.C.
Adanac Moly

high 0.25 grade at Trout Lake, 60 kilometres southeast of Revelstoke. The Ruby Creek property of Adanac Mines is a large low grade moly deposit twenty-two kilometres northeast of Atlin that was hampered initially by lack of a power supply but opened in late 2006. Proven tonnes for National Instrument guidelines are at 38.9 million with a grade of 0.077% molybdenum.

One huge project that has taken more than forty years to realize development and construction is Galore Creek, located in the mountains in the Stikine Gold Belt of northwestern B.C., less than an hour northwest of Eskay creek. The porphyry-related deposit underwent initial work by Rio Tinto and Anglo American before Novagold took control. One of many drawbacks to development over four decades was a lack of power. The present company has purchased a run-of-river hydroelectric plant to power an operation which has a measured and indicated 2.1 million ounces of gold, 24.5 million ounces of silver and 1.9 billion pounds of copper. Mine building began in the spring of 2007, and by summer Teck Cominco announced a major investment in the project which has a hefty $2 billion price tag. From a feasibility study showing a mine life of 20 years, the planned 65,000 tonnes per day mill should put through 420 million tonnes of copper and 400,000 ounces of gold in the first five years of pay back.

The claim stake is an indication of exploration activity
YGC Resources

Sherwood Copper's Minto Mine is the next Yukon copper producer. Production began in 2007 and the mill has a capacity of 2,400 tonnes per day. Located 240 kilometres north of Whitehorse, the measured and indicated categories of the property are an estimated 9.06 million tonnes at a grade of 1.78% copper, 0.62 grams per tonne gold and 7.3 grams per tonne silver.

One great porphyry-related operation in eastern Canada, Gaspé Copper, operated at the town created for it, Murdochville, Quebec, from 1951 to 1999. Originally prospected in 1921, it finally commenced production in 1955 as an open pit. There was a veritable hill of ore which became grandly known as Noranda's Mount Copper. The cost for infrastructure was huge due to the isolation of the area and development required a fifty-one-kilometre hydroelectric transmission cable, part of which ran under the St. Lawrence River. Original estimates were 150 million tonnes of copper grading 1.3%. The smelter provided 97.5% copper anodes and worked on imported ore when the open pit deposit was depleted in 1982.Underground mining from the higher grade zone, 2.6% copper, kept Murdochville alive until 1997. The community was a model town in a beautiful area, but a prolonged strike by a militant union in 1957 resulted in violence and some deaths. Today Murdochville, in addition to tourism and its copper interpretive centre, is working on wind generation as a source of employment. Bill Brissenden, former general manager for Noranda at Murdochville, is in his mid-nineties and still going strong, developing a mine in Mexico. He maintains that there still remains a huge reserve of potentially economic grade at the Murdochville site, which could be developed. The Brissenden family is still exploring for copper on the Gaspé and hopes to revisit mining for copper there.

David Barr with drill core, Galore Creek, B.C., 1963 *D.Barr*

Gordon Davis, plane table mapping, Galore Creek, 1960 *D. Barr*

They sometimes call Niobium by the name Columbium. Anyone in North America requiring the rare, shiny gray metal which is used to improve the mechanical properties of steel and make it more corrosion resistant would find it in Quebec. The source is at Cambior's Niobec mine in Simard Township, fifteen kilometres northwest of Chicoutimi in an intrusive formation. First discovered in 1967, the deposit has been underground-mined since 1976 and the property is the third largest producer of the metallic mineral in the world. The plant has been a good earner for Cambior.

Tungsten is an extremely hard, corrosion-resistant and environmentally-benign metal which may be found

Miner Jim Levesque, Gaspe Copper, Murdochville, Quebec
Noranda

in veins and lode deposits within granite. In some parts of the world it is called Wolframite. The demand for the metal is huge as it is used in super alloys due to its high melting point. North American Tungsten has its CanTung Mine southwest of Whitehorse, Yukon. The deposit was discovered in 1954 by a prospector looking for copper who let his claims lapse because copper mining had not been economical in that location. There have been two closures due to low prices over the mine's life but, reorganized and well funded, it reopened in 2005 and enjoys the enviable position of being the western world's largest supplier of tungsten. In 2006 the operation mined 339,743 short tonnes of 1.09% tungsten ore and should average 400,000 tonnes annually.

Our Best Friend

Kimberlites with Diamonds

Cloak room girl: "Goodness, what beautiful diamonds."
Mae West: "Goodness had nothing to do with it, dearie."

From *Night After Night*

With the discovery of diamonds in the kimberlite bodies of the Lac de Gras district in the N.W.T., Canada emerged as a major diamond producer, challenging South Africa, Botswana, Australia and Russia in both quality and quantity of diamond production. Kimberlite, a rare rock type, is a member of a family of rocks known as 'alkaline intrusions.' These rocks are somewhat closer in parentage to ultramafic rocks, in that they were produced from melting at depths of 300 to 650 kilometres, well into the earth's mantle. This melting occurred under extremely high pressure. Kimberlite forms only where the earth's crust is exceptionally thick, and this occues only under the oldest parts of the earth, the Archean cratons. Canada is blessed with one of the largest and thickest Archean cratons in the world, so it was only a matter of time and the application of a scientific breakthrough before kimberlite was found. The kimberlite melts contain an exceptionally high amount of water and CO_2, and thus their crystallized products are commonly rich in mica and carbonate minerals. As the melt rises, the water and CO_2 form a gas phase, helping to propel the melts upwards towards the earth's surface at near-supersonic speeds. Most of these melts erupted into kimberlitic volcanoes, which are rarely preserved because they are easily weathered and removed by glaciations. In Canada, one such volcanic field is well preserved, at Fort a la Corne in Saskatchewan, but in the Lac de Gras area, only the roots, or feeder pipes to the kimberlitic volcanoes, remain — hence the name 'kimberlite pipes.'

Kimberlite's parent melt actually contains no diamonds; it acts only as an elevator, collecting diamonds from a zone in the mantle about 150 kilometres below the earth's surface as it passes upward. Here the subducted ocean slab that started its ascent thousands of kilometres away is now a rock that has had partial melt removed from it, been exceptionally compressed and completely recrystallized into a rarely observed rock type called eclogite. Diamonds are formed by recrystallizing carbon at these high pressures and are picked up by the kimberlite as it passes through the diamond-bearing mantle on its way to surface. The kimberlite also picks up pieces of the ultramafic rocks that occur in this part of the mantle. During the ascent process, the fragments of eclogite and ultramafic rocks are abraided in what might be considered nature's ball mill. Diamonds, along with other hard, resistant minerals, are freed from their host eclogite and ultramafic rocks. If the rising melt does not come up fast enough, the diamonds transform back to their original form of carbon or graphite.

The scientific breakthrough that enabled the discovery of diamonds in Canada was the observation by John Gurney, professor of geology at Cape Town University, South Africa, that the rocks associated with the diamond-bearing zone, elogite and the ultramafic rocks, contain garnets with a unique, chrome-rich and calcium poor composition. Garnet and associated chrome-bearing iron-titanium oxide material called ilmenite are moderately common minerals and are highly resistant to weathering. Charles Fipke and Stewart Blusson, the geologists responsible for the discovery of the first diamonds in Canada, used this knowledge to sample the glacial materials in large areas of the N.W.T., and on the discovery of garnets with this specific composition, focused their efforts in the Lac de Gras area. Using a combination of surface prospecting and geophysics, the two men discovered numerous kimberlite pipes, many with economically recoverable diamonds.

Kimberlite fields continue to be discovered throughout the Canadian Shield, including parts of the Shield that are now covered by later sedimentary rocks. De Beers will soon have its new mine in production west of the Attawapiskat area, located deep in the Hudson's Bay Lowlands of Northern Ontario.

While on the subject of kimberlite, the importance of another closely related igneous rock, carbonate, should be mentioned. Carbonatite bodies are almost as rare as those composed of kimberlite and are formed deep in the mantle by similar high-pressure melting processes. They and their closely associated alkaline intrusive rocks form circular bodies, much like kimberlite pipes. One important difference, however, is that many of these bodies were injected into fault structures associated with continental rifts. In other words, those areas of the crust where the earth attempted to split apart to make new ocean floor, but only partially succeeded. Volcanism and/or intrusions were abundant in many of these areas, and include carbonatite and related alkaline bodies. Two examples are the great Midcontinent Rift, formed in the mid-Proterozoic, which is an arcuate inverted 'U' that starts in Iowa, swings northeast under Lake Superior, and turns right at Nipigon, Ontario, then right again near Marathon, Ontario, and heads out under Sault Ste. Marie, ending in Michigan. Carbonatite bodies occur near the flexures on the northern side of this rift. The rift system now occupied by the St. Lawrence, Saguenay and Ottawa River systems has carbonatite bodies associated with it at Oka, Quebec. This rift formed at the beginning of the opening of the Atlantic Ocean. Carbonatite bodies and associated alkaline intrusions range in age from Archean (Lac Shortt, Quebec) to the present (in the East African Rift).They are an important source of niobium (Oka and St. Honoré carbonatite intrusions in Quebec), chemical grade phosphate, rare earth elements, and gold (Lac Shortt).

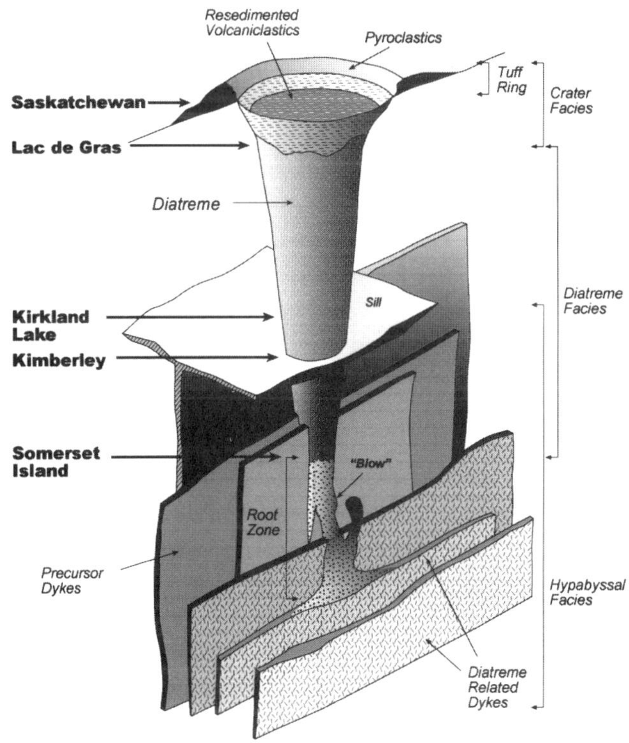

"Diamonds are nothing more than chunks of coal that stuck to their jobs," was the way financier Malcolm Forbes summed up the most precious stone in the world. People perceive them to be rare and yet they are one of the most common gemstones on earth. The rarity comes from challenges in exploration and extraction from the ground. The hardest natural substance in the universe topping out at the high ten on Mohr's 1 to 10 scale of hardness, even the name, which comes from adamantine or unconquerable, denotes a formidable quality. That conviction has been entrenched in the public mind long before De Beer's priceless slogan, "A Diamond is Forever," was launched in 1946.

One reason all the talk about diamonds is extravagant is that they are the oldest objects a person can own. Outshining all other gems with a high refractive index of 2.42, they were created billions of years ago. Every diamond ever found made a quick trip to surface, carried upward in a volcanic explosion. In gauging these most precious stones, formerly one looked at the four 'C's: clarity, colour, carat and cut. Since the last decade of the twentieth century, another 'C' has been added to the list — Canadian.

Diamonds of commercial worth are found in kimberlites and lamprophyre-ultramafic alkaline or igneous rocks with low silica content. These igneous rocks come from deep in the earth and along with heavy minerals and sometimes diamonds, hitch a ride with these mediums as they erupt. The heavy minerals that survive the fierce rush to surface are called indicators and evidence of their presence heightens the search for their precious associates, diamonds. Some of the better known indicator minerals are olivine, garnet, chromite, ilmenite and a variety of crystals with exotic names. Diamond hunting is difficult because kimberlite outcrops are rare, due to the fact that the rock is easily eroded; often a chunk of the stuff will crumble in the hand.

Canadian Kimberlite Emplacement
Queen's University

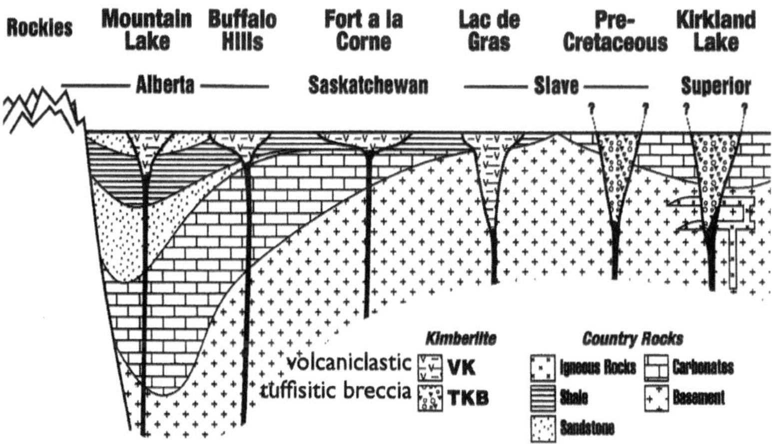

The big mining news in the eighties was of the gold at Hemlo, but in that decade two men were searching for a much more elusive quarry. Veteran prospector Chuck Fipke and geologist Dr. Stu Blusson spent all they had and all they could borrow to finance a quest for diamonds. Diamond exploration in Canada had been ongoing since the early sixties, but it was not until this prospecting team worked out a more systematic approach that there was a real possibility of success. One day, flying in Blusson's tiny, rough-running Hughes helicopter along the edge of the Campbell Ice Field on the western boundary of Banff National Park, they rode the updrafts and eventually spotted a round green patch of rock intruding the Paleozoic limestone. The 'copter path became erratic for a few moments as Blusson shouted, "There's the pipe!" The outcropping was only forty metres in diameter but they landed atop it, grabbed some samples and took off. By the end of the day they had found twenty-six pipes. Each was original, splashed with colour according to the particular minerals metamorphosed on surface: green serpentine; red hematite so rich in iron and limonite; and a nice yellowish hydrous ferric oxide. As it happened, this was one promising area in the hunt, though not the final location.

This exciting day was only a precursor to others doing winter laboratory work plotting the trail of indicator minerals in grab samples, which were obtained from an area extending through 520,000 square kilometres. The indicator arrows finally worked their way north over the glacial divide 320 kilometres northeast of Yellowknife. The place was in the Lac de Gras area of the Northwest Territories. Samples taken at the base of an esker provided 1,500 rare chrome diopsides, a white to light green semi-precious stone, and 6,000 pyrope garnets, a very deep red variety. Ten years of searching had paid off.

Canadian kimberlites compared to South African kimberlite model
Queen's University

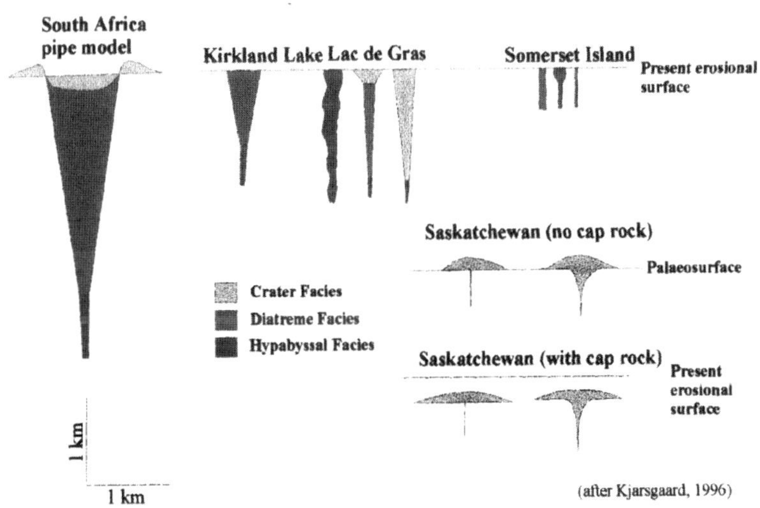

De Beers and Kennicott, a member of the Rio Tinto stable, are very experienced in diamond exploration. Both companies follow similar procedures in exploration. The first year is involved in soil sampling, usually by geologists traveling by helicopter to visit sites eight to eighteen kilometres apart. Soils, glacial till and stream sediments are all used to provide a wide range of samples. Following the search for indicator minerals, LANDSAT satellite and air photo interpretation can lead the exploration into a second year. If the desired formations are found, further exploration will reveal how big the formation is and if it contains diamonds. Years three and four will be taken up first with surface geophysics. This activity will indicate promising drill sites and guide decisions to drill in various locations. Years five and six is when the exploration proceeds to bulk sampling, over 100 tonnes at a time, and both large diameter drilling and often trenching take place.

De Beers spends more than $25 million U.S. a year on diamond exploration in Canada and has even talked of using a blimp to cover the ground in a slow-moving but more comprehensive manner. The total expenditure for all diamond exploration north of the 49th parallel is more than four times that expended by de Beers. In all cases of early exploration, the heavy mineral fraction of the samples is inspected grain by grain to distinguish the species and number of the indicators and determine if diamonds are present in the mix. Transport mechanisms and directions are computed to trace these minerals back to the source area. Analysis of the grains reveals the conditions under which they are formed and if they coincide with those required for diamond formation. Once the carrot- or pipe-shaped bodies are located, the distinctive magnetic and conductive properties they possess may be measured. The geophysical work on the near surface may be done by air or on the ground. Drilling is really the final step in the up to six-year exploration process. Then comes the final test. Hot caustic soda will dissolve almost all minerals from the samples except diamonds. The process of evaluation is concluded when the number of diamonds per kilogram is estimated, as well as the carats per tonne and value in dollars per carat.

Diamond locations in the Northwest Territories

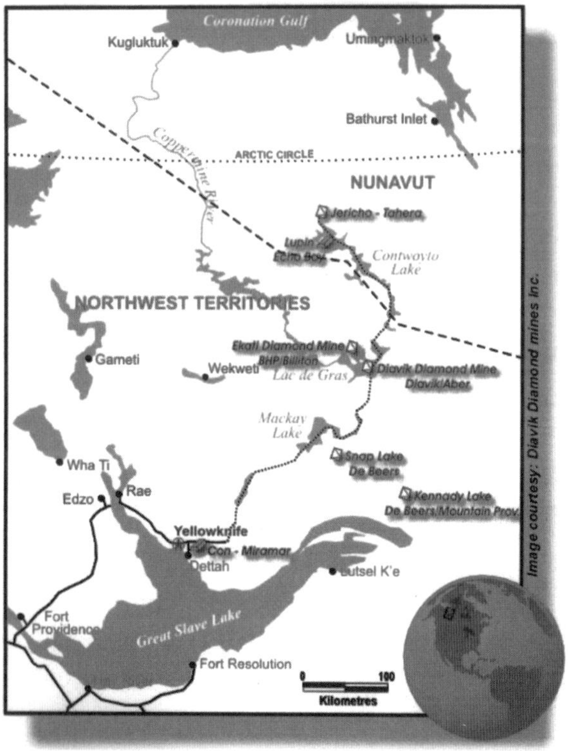

Fipke and Blusson found the source of the diamonds that they had sought for a decade in 1991. The Dogrib and Dene people have a name for the area round Lac de Gras, 300 kilometres northeast of Yellowknife. The location was called e'kati or 'fat lake' and refers to the whiter quartz veins seen there. The streaked rocks are not unlike caribou fat, a commodity prized long before diamonds brought the remote location into prominence. The diamond hunters formed their own company, DiaMet Minerals and utilized a new aerial sonar technology to pinpoint the host kimberlite pipe. All this took time, and the partners spent all available funds on drilling and workers. In the midst of it all one helpful employee showed Fipke several chunks of a crystal rock found just lying on the ground. Too bad they were only large quartz crystals!

Logistical challenges were huge in the building of the Diavick Mine in the N.W.T.

Jiri Herman, Diavik D.M.

Then one day Chuck Fipke's son Mark saw a tiny emerald green stone lying on the glacial till. The one centimeter-size soft crystal was a chrome diopside from deep in the earth, likely torn from the surface of the kimberlite pipe. The searchers drilled the little unnamed lake which was in the centre of a peninsula called Point de Misère and gave it their own name, Point Lake. After they had collected 1,200 samples, an authority examined the data and felt the grade would be better than sixty carats per one hundred metric tonnes. When the ore was concentrated, Fipke found eighty-one tiny diamonds. In the age-old tradition of realistic prospectors, the partners realized early that a major player with deep pockets would be necessary for further progress in the venture. They teamed up with Australian miner BHP Billiton and received 10% each of what would become The Etaki Mine for their well-deserved share. Once the news of the staking became public, the rush was on; fifteen mining companies sent field workers to stake the ground.

Since that diamond rush took place in 1991, almost a century after the Klondike gold stampede, Canada has been ranked among the top three diamond producers in the world in terms of value delivered. In the first five years of operations, Canadian diamond mines produced 13.8 million carats worth $2.8 billion. Statistics Canada put this in an uncharacteristically colourful way. The federal keeper of all things factual said that this represents 1.5 kilograms of 'ice' each day for five years with each shipped 'bag' worth $1.5 million. Remember: kimberlite pipes do not necessarily contain diamonds. By 2002 over 300 pipes had been found in the Northwest Territories alone, but only fifteen to twenty-five offer promise of economic potential.

The Etkati Diamond Mine owned by BHP Billiton and DiaMet took seven years from discovery to mine production. The original prospectors knew they had a winner when the mining process brought up several large diamonds, one a handsome 182 carats. This news was the payoff for an almost $900 million project which included trucking 40 million kilograms of material over an ice road from Yellowknife to the mine site 475 kilometres away.

The discovery of the Diavik deposit is the sort of story that has investors reaching for their cheque books. In 1992 Eira Thomas was a young geologist leading an exploration crew east of what became the Etkai Mine. The area was around Lac de Gras and the company was Aber Resources, and one of the directors was her mining engineer father Grenville Thomas. The twenty-four-year-old geologist, who had only received her professional designation two years before, had a crew drilling for days without success in the middle of a big lake, which was starting to melt. The drillers were agitating to pack up the rig when Ms. Thomas looked at the latest sample and discovered a perfect 2.5 carat diamond. As ever, a partner with large resources was needed to ensure financing of the lengthy evaluation and development process for this property. Rio Tinto came in at 60% with Aber holding the 40% balance of the property. Permitting, bulk sampling, and construction (which included the first of three dykes to enable dewatering the lake) located on an island, took from 1992 to 2003, when the first diamond production began after an outlay of $1.3 billion.

Diavik infrastructure
Diavik D.M.

In the first five months of operation Canada's second diamond mine produced one million carats. At peak the operation should produce 10 million carats annually. In 2007 the company had exceeded $1 billion in business with Aboriginal firms. As for the geologist, she became Vice President of Exploration. Her story continues in another diamond quest.

Up Here Publishing

Nunavut's first diamond mine, and the third in Canada, was opened in August 2006, eight years after the diamond potential of its claims was realized.

A smaller property than the Etkai and Diavik mines, it has an expected mine life of eight years, but Tahera Diamond Corp's Jericho Mine enjoys the distinction of having had prestigious New York jewelers Tiffany & Co. purchase a significant portion of its output and promote the diamonds. The mine had a much lower capital cost than its predecessors, at $104 million, but, after opening, had to make a strategic alliance with Teck Cominco to obtain both financial and technical backing. The mine, 420 kilometres north east of Yellowknife, is just inside the Nunavut border. The diamonds come from a land-based pipe with three phases or lobes. Average annual production is hoped to be in the range of 100,000 carats; so far, financial results have been disappointing.

De Beers' Snap Lake project is 220 kilometres north east of Yellowknife and will be the first entirely underground diamond mine in Canada. The mineable resource of 22.8 million tonnes should ensure a mine life of twenty years from its October 2007 start up. Since De Beers acquired the ground in 2000, capital costs should top out at $975 million. The world's major diamond player also has the first diamond mine in Ontario, the Victor Mine in the James Bay lowlands, ninety kilometres west of Attawapiskat. Investment here is also huge, at $982 million, and during the estimated twelve years of mine life, beginning in 2008, the plant should give up close to 600,000 carats a year. De Beers has 51% of the Gahcho Kue property, 80 kilometres south east of Snap Lake, and will be the operator of the mine, which is scheduled to open in 2012.

This junior company uses a catchy image to promote its business.
Spider Resources

There are scores of companies actively exploring for diamonds in Canada, and the search is spread over five provinces and two territories. The sparklers attract investors from venture companies. The area of concentration in Quebec is the central part of the province north of the Otish Mountains. In Ontario there has always been promising evidence of diamonds as they have been found in glacial till in southern Ontario, thought to have been displaced from the north. Apart from staking in the vicinity of the Victor Mine, there is interest in the Highway 11 corridor between North Bay and Kirkland Lake, especially near New Liskeard and Temagami. Wawa is also an area busy in the hunt for diamonds. Manitoba is seeing activity west of Churchill in the Hudson Bay Lowlands and Seal River. The Fort à la Corne area fifty kilometres northeast of Prince Albert, Saskatchewan, has several diamondiferous clusters found in large, classic 'champagne glass' formations found beneath approximately 100 metres of glacial material.

A 24 inch drill bit is lowered into the casing.
Peregrine Diamonds

Alberta has several locations of interest to diamond prospecting firms. New areas of interest include the foothills in the northeast of the province. Fifteen firms are actively exploring for diamonds in the Northwest Territories, spurred on by the successes at Ekati, Diavik and Snap Lake properties. Most of the interest is in the Slave Region or craton. Eira Thomas who controls Stornoway and newly acquired Ashton Mining has received much attention due to her work at Diavik. In Nunavut the presence of the Tahera Mine has acted like an exploration magnet for junior companies. One of the most promising is Peregrine Diamonds which has had a highly successful bulk sampling result in its DO-27 Lac de Gras kimberlite pipe.

The hunt for diamonds in the far north of Canada is hot. De Beers offered a career opportunity recently for a geology specialist that would have been unheard of in this country a couple of decades ago. The exploration division was looking for a volcanologist at the Masters or PhD graduate level. Aboriginal and northern workers have particularly benefited from the exploration boom now in progress. De Beers is investing large sums in the training and education of high school graduates in the Northwest Territories. Diamond cutting is the subject of an up to forty-two-week course at Aurora College. Aber Diamond Corporation has actually gone so far as to buy out luxury jewellery retailer Harry Winston as a vehicle for its diamond sales.

Traveling the winter road to Snap Lake *De Beers*

The term 'blood diamonds' became current in the nineties. Sometimes called 'conflict' diamonds, these gems were stones from the African continent which were used to finance rogue armies and terrorist organizations. In 2002 the Kimberly Process for tracking and certifying diamonds was set up by thirty-seven diamond trading and production companies. Purchasers are guaranteed that the origin of their diamonds is free of illicit trade. This process has been most beneficial to Canada as the diamonds originating in its north country are free of any association with such conflicts and thus even more desirable. The story of diamonds in Canada is still only in the beginning stages.

Polished diamonds from the Jericho Mine as presented by Tiffany's of New York
Tahera Diamond Corp.

Faults and Fissures

Vein Deposits

***The gold, it doesn't rot, it doesn't eat;
it just sits there and waits.***

Paul Penna

The discovery of silver and gold vein deposits marked the start of Canada's mining legacy. The discovery of gold at Kirkland Lake and Timmins and silver in Cobalt and near Thunder Bay set the stage for the development of these parts of Canada's hinterland and founded the development of a mining culture that continues today. Almost all of the deposits types discussed in this review contain veins. In many, such as the stratabound deposits, these are not the principle mode of ore occurrence, although they may provide very high-grade 'sweeteners' to add to the more stratiform mineralization. Some, such as the porphyry and related scarn group, are comprised of 50% to 100% veins, but these have a close association with major intrusions. Two important deposit types, however, lode or vein gold deposits and Cobalt-type silver deposits are predominantly contained in veins which occur in fault structures that are usually related to major crustal-scale deformational events and which have a somewhat more tenuous or no relationship to intrusions or processes that formed the stratiform group of deposits.

Lode Gold

Discovery of gold deposits in the Canadian Shield together with the discovery of Cobalt-type silver and copper-zinc VMS deposits at Cobalt, Ontario, and Noranda, Quebec, underpinned the development of the Canadian mining industry as a major economic contributor. In contrast to the VMS deposits, many of which were discovered in the period after World War II by geophysical methods, virtually all of the gold deposits were found by prospectors or by geologists pursuing gold along well-developed corridors in the Timmins, Kirkland Lake and Val D'Or camps using the drill and trench. In British Columbia, placer miners, recovering gold from streams by sluicing and panning, began to realize that these had a local bedrock source, and soon traced the gold to nearby vein deposits.

Early in the discovery of the vein deposits in Ontario and Quebec, it was recognized that the deposits occurred in distinct linear trends in each of these camps. These became known as the 'breaks'; Kirkland Lake, Larder Lake, Destor Porcupine and Cadillac Malartic breaks became everyday terms in the exploration community. Eventually similar breaks were noted in the Beardmore-Geraldton, Red Lake and Yellowknife camps as well. Typical of these breaks, the Destor Porcupine Break has all of the deposits from Timmins to Duparquet, a distance of over 200 kilometres, with smaller deposits extending for 60 kilometres within an eight-kilometre-wide corridor. Although many discoveries have been made in new areas, most of them also have the same spatial relationship to a major geological 'break.' These observations lead us to a series of questions: Why are the deposits so closely related to these breaks? Why do the quartz and carbonate-filled veins that host the gold seem to cut across virtually all rock types? Why are some areas along these breaks replete with major producing mines and others seem to be devoid of much gold? Why are the vein systems in Ontario and Quebec so rich, whereas in similar geological settings in northern Manitoba and Saskatchewan, only a few significant mines have been found? Are the gold deposits in the Cordilleran regions of B.C. and the Yukon the same as those in the Shield? Finally, super-giant districts elsewhere on earth, such as the Witwatersrand district of South Africa (by far the most productive district in the world), the huge deposits in the Carlin and other districts of Nevada, and the major gold districts in the Pacific islands and western parts of the Pacific 'Ring of Fire' do not seem to have equivalents in Canada — or do they? These questions cannot be answered here but at least the geological characteristics in Canada can be examined.

The 'main breaks' typically juxtapose geological domains of different ages, usually of different rock types, and, in some cases, different tectonic settings. In many areas there are sedimentary dominated terrains on one side, volcanic-dominated on the other. There are complex fault zones, which represent zones of collision between terrains much as the continents are colliding today, except there is scant evidence that today's collisional zones existed during the Precambrian period. These collisional zones were, in some cases, oblique; in many, they recoiled to form local basins that filled with terrestrial sediments; and, at Kirkland Lake, because of volcanic products, they became major zones where fluids could escape from very deep in the crust. These micro-continental blocks that were colliding were totally crystalline, and probably already deformed. Although it is not entirely clear how these fluids were generated, there is consensus that they formed through normal progressive heating and dewatering of crustal materials with depth, but at major suture zones they found a way to escape. Some melting of the crust also occurred in these collisional zones, and small bodies of granite (generally called quartz-feldspar porphyry because of their texture), a common associate of many gold districts.

The fluids contained much silica, the parent element for quartz, and considerable sulphur and CO_2, along with gold (only a few tenths of a part per billion) and a small amount of iron and base metals. As they rose to within a few kilometres of the earth's surface, they began to boil and discharge gases, mainly water vapour and CO_2. These vapours reacted with the rocks surrounding big faults, causing them to become carbonatized, imparting a tan to pale red colour so prominent around many gold districts. The pressure of these fluids exceeded the abilities of the rock to contain them, and massive fracturing occurred, into which the silica, some carbonate, a small amount of pyrite and most of the gold precipitated. These were violent, multi-stage events, causing the rocks to crack, seal with quartz and gold, and crack and seal again and again. This structural disturbance did not happen in the main faults but in subsidiary structures that were parallel or at a slight angle to them. In areas where brittle and ductile rocks were adjacent, the brittle rocks fractured and opened up for spectacular gold quartz deposition, whereas the ductile rocks flowed or split into very thin plates of rock that together formed 'shear zones.' (The Red Lake Mine is a good example of gold at a ductility contrast.) In some areas, gold precipitation was aided by reaction with iron-rich rocks, such as the iron formation noted above and present in districts such as Geraldton and Pickle Lake, Ontario, and Lupin in the N.W.T. The combination of the right structural regime, the right host rocks and a very deep penetrating collisional zone set up these areas to be the really big gold districts. Why are these fault systems not continually mineralized? We find the answer in the fact that some parts of the fault resisted the necessary dilation that enabled deposition, whereas others remained under compression and were too tight to allow for the explosive dilation that set up the vein-forming process. Why didn't this work as well in Manitoba and Saskatchewan? It probably did, but these areas seem to be at a lower structural level, or alternatively, the dewatering process happened before any sort of collision and suture forming event and the fluids were dispersed or reincorporated into the crust.

Epithermal Vein Systems

Many of the gold deposits in post-Precambrian sequences and even some deposits in the Shield formed by a set of different processes that have close affinities to both the porphyry-copper and less directly to the VMS systems mentioned previously. These deposits, which are more abundant in the Yukon and British Columbia, are generally termed 'epithermal deposits.' The term 'epithermal' means that they formed at or near the earth's surface, usually at less than 1500 metres, but typically only a few hundred metres, from water that was heated as much as 400C as a consequence of magnetic activity. Most epithermal deposits formed in ocean-continent collision areas that were at least partially above water, either as an island arc such as Japan, or near the subducted margin of a continent such as South America. Thus they tend to be in relatively young rocks, simply because they are best-preserved in areas which had not suffered the depth of erosion that is common in older terrains such as the Shield or the Appalachians. Some excellent examples remain in these older areas but they are much less common. Most are associated with sub-aerial volcanism and are in the volcanic rocks that overlie major intrusions.

Epithermal deposits come in two distinct types. One, termed 'high sulfidation epithermal' type, is associated with abundant sulphide minerals, a silica sinter cap, and abundant quartz alteration and veining, along with an exceptional amount of potassic alteration, including alunite, a potassium sulfate. Other sulfate minerals such as anhydrite (Ca sulfate) and barite (Ba sulfate) may also be present. Lead and zinc sulfas may be abundant as well. This type may form above a porphyry system, and may be transitional into the upper parts of porphyry copper systems. The fluids that form them are direct emanations from the magma chamber. Deposition occurs at the boiling point of these fluids, usually at several hundred metres below the paleosurface. Active systems manifest themselves as hot springs on the surface, which are usually accompanied by large volumes of silica also known as 'tufa.' Some are rich in tellurium-bearing minerals; some have significant base metals and high silver content. The veins are banded and complex. The upper part is sometimes rich in elements that are easily transported as volatiles such as antimony and arsenic sulphides (stibnite, realgar and orpiment) and mercury sulphides (cinnabar).

The second set of this type are the 'low sulphide epithermal' deposits. These are associated with veins rich in a special form of potassium feldspar called 'adularia' and the veins are surrounded by sericitized and silicified wall rocks. They contain abundant silver and some are primarily silver deposits. The mineralized veins are banded and zoned, but in the upper, non-mineralized parts of these deposits, chalcedony predominates. Base metals are typically absent, but arsenic minerals are common. Both types have extensive zones above the best mineralization where acidic steam virtually destroyed the original rocks. These intensely altered areas are composed of aluminous clays, usually kaolin, mixed with silica. Many of the low sulphide types are associated with massive areas of exploded rock, hydrothermal breccias, through which steam-heated vapour passed. The near surface parts of some low-sulphidation deposits contained so much mercury that they were mined for it, as well as abundant antimony. Some districts contain deposits that are intermediate between the two just described. In older sequences, there are enigmatic deposits, such as those in the Hemlo camp of the Canadian Shield, and in several in the Kirkland Lake area that have many characteristics of high-sulphidation type deposits. The Kirkland Lake deposits are in late sub-aerial volcanic rocks and contain abundant telluride minerals. At Hemlo, sulphides, including molybdenite, are common, and barite is a major mineral. There is a question as to whether these are formed by epithermal or porphyry processes, but recent research has not reached a definitive answer.

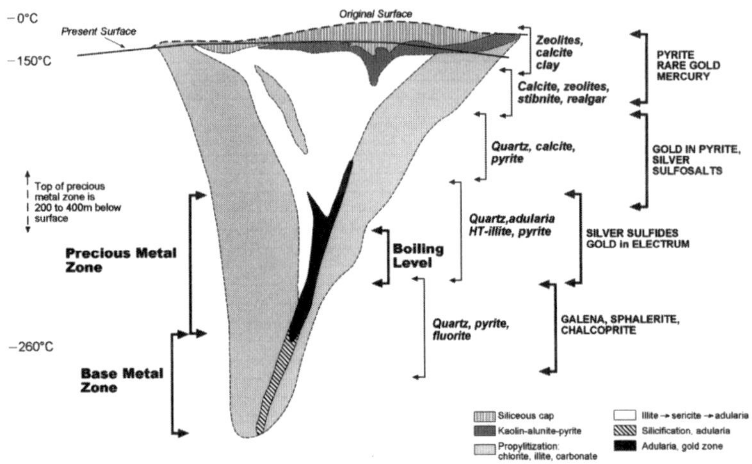

Art Buchwald, pointing to the steady demand for gold, once suggested that the United States should end its love affair with the precious metal and turn to a junked-car standard. These would ease monetary woes and clean up the landscape. Old wrecks would be crushed to shoe box size by the Federal Reserve Bank while gold was dumped in the nearest river. Since Americans junk cars faster than any other nation, he reasoned, their reserves would be the world's largest. The humorist's novel approach to fiscal reform ignored one basic fact: the value of gold goes beyond the price dictated by the world market. Gold would be attractive even if it cost nothing. As soon as the gold was dumped in the river, hordes of people would be waiting to retrieve it and carry the stuff away. After all, the lustrous yellow metal is inert, never loses its value and only appreciates. Investors thought so in 2007. The *Northern Miner* had the headline, "Hordes Hoarding Gold," on a story about investors who snapped up 44.5 million ounces in 2006 and for the first time held more of the yellow metal than central banks.

The Siscoe Mine, Val D'Or,QC

Denis Newman

Quebec enjoys the fruits of more than $200 million annually spent on exploration in the province. The province is considered most attractive to exploration companies, due to its cash rebate incentive program for exploration projects and its more liberal provincial tax reduction for Quebec flow-through investors. Numerous new mines are coming on stream, so it is obvious that the government incentives are paying off. Camps such as Val d'Or, have repaid investment over the years with 13,180,000 ounces of gold refined to date. Current gold operations are a mix of old and new and there are several advance projects on the horizon. Wesdome's Kiena Mine deposit was discovered just prior to World War I but languished for many years. The plant is located ten kilometres west of Val d'Or. It sits on Parker Island and is accessed by causeway across the southwest arm of Lac de Martigny. The ground is carbonate-quality stockworks occurring in veins. The property has had several owners including Falconbridge, Campbell Red Lake and McWatters, who ran it to 2002 before going bankrupt. From 1981 to 2000 the mine produced 1.56 million ounces grading 4.75 grams to the tonne. The mine was closed until bought by Wesdome and reopened in 2006.The new owner carried out exploration work and established reserves at 1.4 million tonnes at 5.2 grams per tonne grade. The plant was easy to put into production as it was on standby and the pumps were still working. The mill has a daily capacity of 2,300 tonnes and 52,000 ounces of gold should be delivered in 2007. Quebec miners are aggressive and even small properties are being taken up. Richmont Mines began producing at its East Amphi Mine in 2006 and from measured resources of 1.39 million tonnes averaging 5.4 grams gold per tonne, it should commence operation with 25,000 ounces recovered in its first year.

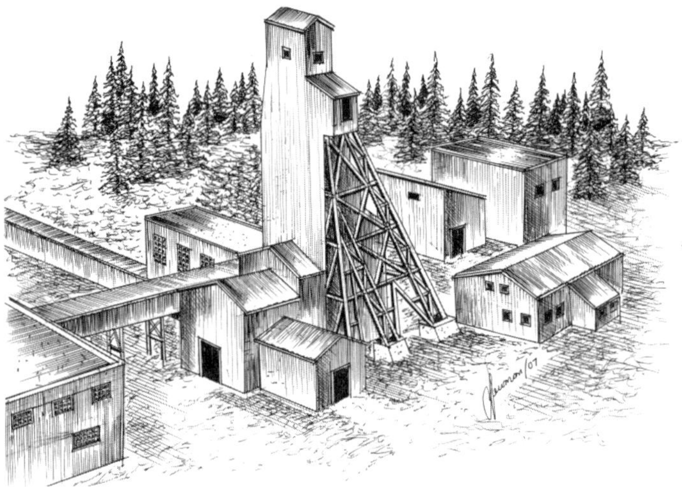

The Sigma Mine, Val D'Or, QC,1975
Denis Newman

The Doyon mine is 40 kilometres east of Rouyn-Noranda on the Cadillac portion of the Abitibi region. The gold there is found in veins and pyritized schist. The mine commenced operation in 1980 and was owned 50% by LAC Minerals and the balance by Soquem, Cambior. In 1986, Iamgold bought Soquem's share and Barrick obtained the LAC portion in 1998. For a quarter century with its open pit and underground mine, Doyon delivered a handsome total of 5 million ounces of gold before Cambior bought out the Barrick share of the property. In 2006 the mine produced 157,000 ounces at 6.4 grams to the tonne grade.

Mines in Quebec seem to play musical chairs and, in this case, Aurizon originally owned the Sleeping Giant Mine, eighty kilometres northwest of Amos. This mine deposit was found by Mattagami Lake Exploration using airborne geophysics in 1976 while searching for VHS deposits. The deposit was discovered to be a vein type in smoky quartz with high sulphide content. Typical of many small mines in the area, there was never more than a couple of years of reserves of ore.

Mining began in 1988 but ceased by 1991 due to low commodity prices and limited reserves. The mine was reopened in 1993, and the shaft was deepened that year and again in 2004 to 1060 metres, a year before Cambior bought it from Aurizon, undertook aggressive exploration, and successfully located higher grade ore. In 2006 the estimated production at a handsome 11.1 grams to the tonne was 126,000 ounces of gold with Iamgold as the operator.

Aurizon's Casa Berardi Mine, 95 kilometres north of La Sarre in the Abitibi gold belt, gains its name from that of the township, part of a movement to commemorate World War II battles. Inco found and built the property and operated the extraction of the gold-bearing quartz vein system until 1998 when Aurizon acquired the mine. The property was closed and on care and maintenance until the price of gold rose and warranted more underground exploration. The underground workings join two plant locations with a tunnel five-kilometres long. The newly refurbished property reopened in late 2006 and was expected to produce an average annual recovery of 165,000 ounces at 10.1 grams to the tonne in 2007.

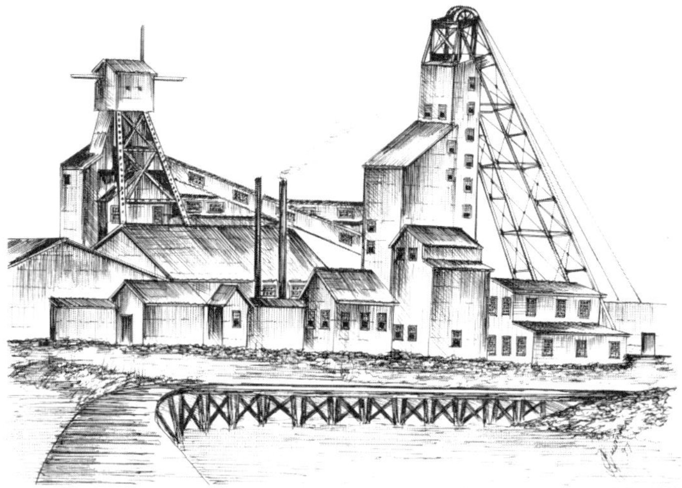

Sir Harry Oakes' Lake Shore Mine, Kirkland Lake, Ontario
Denis Newman

Campbell Resources has the Joe Mann Mine not far from its Copper Rand property in the Chibougamau region, 550 kilometres northwest of Montreal. The deposit is vein copper and gold, and in 2005 the neighbouring Copper Rand mill delivered 29,431 ounces of gold at 0.25 grams per tonne grade along with 897,460 pounds of 0.2% copper. The small property had precious few reserves, but the nearby Corner Bay copper prospect will likely become a mine.

The Sigma and Lamaque mines close by Val d'Or in the Abitibi region have a long and profitable history. The Sigma mine worked from 1937, five years after discovery, to 2000 under three different owners and only closed due to a bankruptcy matter. The adjacent Lamacque mine worked from 1935 to 1985 on the same vein system. Century Mining acquired the Sigma mine in 2004 and recommenced mining the next year. This firm introduced the super pit open mine concept which renders low grade mines more profitable, as at the super pit at the century-old Dome property in South Porcupine, Ontario. In 2006 the product amounted to 73,000 ounces at 2.1 grams to the tonne. Adjacent Lamaque mine is currently dormant but recent exploration has increased its mineable reserves. Just west of Val d'Or, Agnico Eagle has its Lapa Mine which holds mineralization in veins in volcanic rocks. Originally mined between 1937 and 1943 to the tune of 50,000 ounces of gold, it closed because of the common problem of obtaining men and material due to other priorities of the war effort. Exploration was done intermittently over the years. The present owner made the decision to proceed, and work commenced on construction in 2004. Proven grade is 10.65 grams per tonne, and the mine was due to start production in 2008 on recovering its estimated 1.1 million ounces of gold. Agnico Eagle has another gold property due to come into production in 2008. The Goldex Mine is on the western outskirts of the Val d'Or area. The latest underground bulk sample returned a grade of 2.8 grams per tonne in ultramafic rocks. Reserves are 21.77 million tonnes grading 2.4 grams gold per tonne, and commencing in 2008, production should be 170,000 ounces of gold per year from large, low cost underground mining.

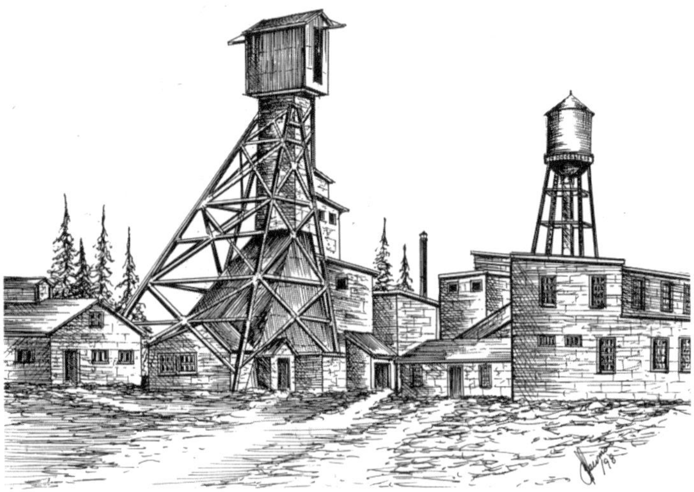

The Macassa Mine, Kirkland Lake, ON, has been operating since 1932.

Denis Newman

Gold mining has come a long way in Ontario since the first property, the Richardson Mine in Eldorado near Madoc, fizzled shortly after its 1867 opening. The scattering of small mines working in northwestern Ontario eked out a few ounces of gold in the early part of the twentieth century. The success of the Cobalt camp gave witness to the Mexican proverb, "It takes a silver mine to make a gold mine," by providing a labour pool and ready financing for the rich gold bonanzas of the Porcupine and Kirkland Lake. The Porcupine-Timmins area produced 67 million ounces of gold from 48 mines between 1910 and 2004. The smaller but richer grade Kirkland Lake camp had an output from twenty-four mines that gave up 42 million ounces between 1917 and 1990.

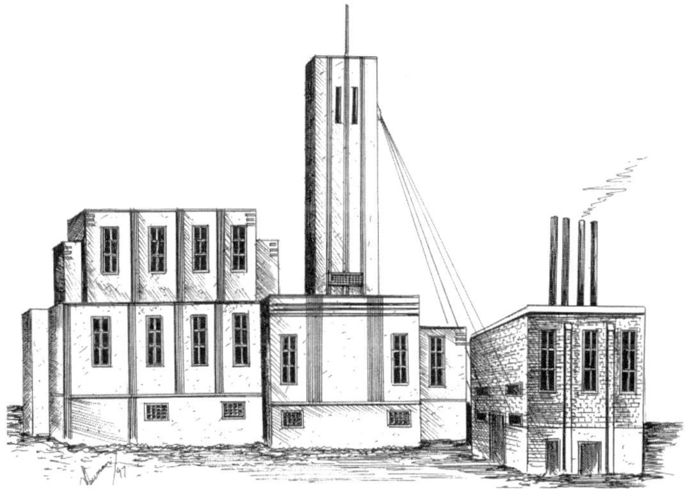

Although long closed, the Hollinger Mine still stands, its art deco- style buildings dominating the Timmins skyline.

Denis Newman

The Kirkland Lake and Timmins camps are in close proximity to each other and are both within the Abitibi, the largest greenstone belt in the world. The major producers in these camps are or were in Kirkland Lake: the Toburn, the Wright Hargreaves, Lake Shore, Teck Hughes Kirkland Lake Gold and Macassa. Farther east are the Dobie-area Upper Canada Mine and the great Kerr Addison Mine at Virginiatown, close by the Quebec border. The Timmins-Porcupine camp is much larger in land area and has had among the larger properties: the Aunor, Coniarum, Dome, Pamour, Hollinger, Hoyle Pond, McIntyre, Owl Creek, Paymaster and Preston. There is a mixture of mineral deposit types along the two camps and they have been placed together for convenience. All are confined to the Porcupine-Dector Fault structure and its splays or subsidiaries.

Three gold mines are operated in Timmins as a joint venture, the principal owners banding together to avoid duplication of resources and to obtain economies of scale. The Dome Mine was founded in 1910 and is the oldest in the camp. The largest and second largest mines in the Timmins camp commenced production shortly after the Dome went into operation. However, the Hollinger and MacIntyre, adjacent to each other and with connecting underground mine workings, ran out of economic ore in the early sixties. The Pamour Mine, in which famed mining pioneer Viola MacMillan had an early interest, commenced operations in 1935, one of many properties which came into

The Young Davidson Mine at Matachewan, ON, is not operational in this view. Note the hoisting cables are not attached.

Northgate Minerals

Profits from the McIntyre Mine, Schumacher, funded the Maple Leaf Gardens and Banting's research on black lung disease.
Brenda Burey

being when President Roosevelt increased and stabilized the price of gold at $35 (U.S.) an ounce. Hoyle Pond was originally a project of Texas Gulf and began operating in 1982. The Dome is located on the outskirts of South Porcupine; Hoyle Pond is five kilometres north of Highway 101, the main road to Timmins; and Pamour is two kilometres south of the Hoyle property. Visitors cannot miss the giant pit as it is right on the highway. The Pamour headframe is gone now, and Highway 101 has been rerouted once to permit more mining. A new direction for the main route to the city has been carved out again to facilitate joint venture mining. The Dome geology offers veins and disseminated sulphides. Pamour is a vein structure and at Hoyle the gold is either native or in a pyrite mix. In 2006 the partners shared 374,921 ounces at an unspecified grade.

The Macassa Mine commenced producing gold in 1932. The last mine on the three miles of gold of the Kirkland Lake camp is almost at the centre of the famed Abitibi greenstone mineral belt. The ore in the camp tends to trend progressively deeper from east to west, and pioneer prospector Harry Oakes (later Sir Harry) was one of those who suggested that gold would only be found at depth. After a succession of owners, the Macassa Mine is now controlled by Kirkland Lake Gold Ltd. which also has the ground of four other former mines including Lake Shore which made the Oakes fortune, and the Wright Hargreaves, whose proceeds enabled Bill Wright to found the *Globe and Mail* newspaper. Located off Highway 11, 240 kilometres north of North Bay, Macassa in its time has had annual production of more than 100,000 ounces but the property was closed due to rock bursts and shaft damage in 1999. The present owner instituted an aggressive exploration effort and in 2005 45,865 ounces at a 0.335 grade were recovered. Reserves had been raised to 927,000 ounces by the close of 2005, and it was expected that output would be closer to 90,000 ounces in 2006. The Macassa story is happier, with exploration potential at depth excellent, compared to the giant Kerr Addison Mine at Virginiatown, close by the Quebec border, which has unknown remaining exploration potential.

Once the largest gold mine in the country, it mined 10,457.441 ounces at a fine 0.259 ounce grade before it was closed after six decades of service following a take-over by a junior that eventually went bankrupt.

A string of small mines have operated east of Matheson on Highway 101. The area has always attracted interest from mining buffs because of the Croesus Mine which operated intermittently from 1915 to 1936 at a handsome 2.8 ounces to the tonne grade giving up 14,859 ounces, mostly in the area of the shaft. The largest producer was the Holt-McDermott mine which operated from 1988 to 2004 with 1,322,332 ounces at a 0.162 ounce grade. This mine was associated with well-known miner Dit Holt. Across the road another Noranda Exploration success, Holloway Mine, was also started up in 1988 and is now owned by Newmont, annually delivering 75,488 ounces at 6.0 grams to the tonne. The company has acquired the ground of its neighbour, the Holt McDermott.

Red Lake mines on the skyline

Carl Huston

There are always small gold mines under development or in production in the area of Wawa, where the more dependable iron was mined for years to feed the Algoma Steel mills in Sault Ste. Marie. Richmont Mines has started to produce at its Island Gold property near Wawa. The big news in 1979 was the staking of three properties some 200 kilometres west on Highway 17 close by Manitouwadge, which finally beat the nearly century-old jinx which seemed to prevent the establishment of a viable ore body at Hemlo near Marathon, inland from the north shore of Lake Superior. Successful prospectors John Larche and Don McKinnon were drawn into the morass of lawsuits that resulted from the discoveries; Lac Minerals lost its property as the result of one action, and original booster Murray Pezim was eventually ousted from his significant role in one of the mines. In the end there were three properties which had resulted in a huge staking rush. Newmont had Golden Giant, originally acquired and built by Noranda, and Barrick and Teck Cominco shared the David Bell and Willams mines, originally owned by International Corona. The geology of the camp on the greenstone belt is stratabound sulphide, gold finely disseminated in volcanic rocks in a very unique and complex ore body. The Golden Giant lasted until 2006 and delivered 170,929 ounces of 10.227 grams to the tonne in the last year. The David Bell mine had 105,143 ounces at 9.6 grams to the tonne grade, and the big Willams mine produced 354,884 ounces at a 2.57 ounces to the tonne grade in 2005. The Williams took care of the output of both joint venture properties with its 10,000 tonne-per-day mill and by 2005 the camp had produced 19 million ounces. The three mines at Hemlo all produced from the same ore body. This resulted in duplication of administration and milling. This waste of effort is unfortunate but unavoidable when the owners are competing for each other's properties.

The Musselwhite Mine located 430 kilometres north of Thunder Bay, accessed by air and by all-weather road from Pickle Crow was discovered by prospectors in the late sixties but it took a quarter of a century for exploration, financing and development of infrastructure. The fly-in-fly-out operation is hosted by banded iron formations with much pyrrhotite and quartz and rarely visible gold. Presently the mine is held by Goldcorp with two-thirds of the ownership, and Kinross has the balance. In 2005 the output was 250,383 ounces of gold at 5.7 grams to the tonne grade. The mine has a large complement of Aboriginal workers and First Nations people also supply air transport and camp facilities.

The Red Lake gold camp was the next big mining event in Ontario after the discovery of the Kirkland Lake camp. The first Red Lake rush in 1926 took place in the dog team era, but within ten years miners were using airplanes. The first production year was 1927 and, by the close of 2005, 22,882,403 ounces of gold averaging a fine 0.438 had been extracted from twenty-eight mines in the vicinity. Mines that have stood out above others with more than a million ounces of gold have been the Campbell, Dickenson (Goldcorp), Madsen, Cochenour-Willans, Mackenzie, Howey and Hasaga.

The Dickensen and Campbell mines were both established after World War II, at Balmertown near Red Lake, 175 kilometres north of the Trans-Canada highway at Vermilion Bay. The two mines within sight of each other shared an ore body along an east-west deformation zone in a greenstone belt with free gold often mixed with sulphide minerals, mainly arsenopyrite. The Campbell was always the richer mine and the Dickenson appeared mined out in 1994 and had suffered four years of strike action. In 1995 under new owner and Goldcorp president Rob McEwan, the company discovered a rich, new deposit below the current mine. Goldcorp had acquired the property and through a highly innovative internet challenge, it threw open all maps and reports of the ore body for inspection and offered prizes to those who could figure out how to proceed in discovering more ore. The mine was rebuilt on the surface with a new shaft and began to produce in 2001. Goldcorp took over the Campbell complex in 2006, and the newly combined Red Lake Mine literally had a golden future with reserves of many millions of ounces. In 2005 the Goldcorp portion had produced 616,400 ounces at a rich 82.0 grams or over two ounces a tonne grade. The Campbell mine delivered 440,000 ounces at 15.0 grams per tonne in the same period and a million ounces a year for the combined operation is now within sight. Recent exploration on the old Gold Eagle property has delineated the down plunge extension of the Cochener-Willans deposit. The present Gold Eagle, formed by combining Southern Star and Exall, has strong potential in its Bruce Channel property, which lies at the intersection of the Red Lake east-west and north-east deformation zones.

The Long Lac Mine

Denis Newman

There were ten gold mines in the Beardmore, Geraldton and Long Lac areas in the district now encouragingly known by provincial designation as Greenstone, no doubt for the Keewatin Greenstones and Temiskaming sediments found there. Three are worthy of note. Gold was found in the area in 1917 but it fell to veteran prospector Bill 'Hard Rock' Smith to draw attention to the emerging camp. Famed geologist Murray Watts worked with Smith at one point as a young man and recalled the character's daily admonition: "We're destined to find the big one today. Let's get going!" Smith found his gold in 1931 and called up his friends Frank MacLeod and Art Cockshutt who made a discovery which took several years to make a mine. Finally the McLeod-Cockshutt began delivering mill feed in 1938 and produced until 1968, giving up 1,475,728 ounces of gold and 101,388 ounces of silver produced from 10,337,229 tons at a 0.14% grade. Prospectors Tom Johnson and Tony Oklend were handsomely rewarded by promoter Joe Errington and Thayer Lindsley as they built Little Long Lac Mine which opened in 1934 and by closure in 1956 had given up, from 11,780,516 tonnes milled, an acceptable 605,499 ounces of gold and 52,750 ounces of silver at an average grade of 0.34 ounces to the tonne. As for Bill Smith's Hard Rock property, found in 1931, it began to operate in 1938 and closed in 1951. By the end of its life the property gave up 269,081 ounces or gold and 9,009 ounces of silver from an average grade of 0.18 ounces to the tonne.

A messy gold pour at Contact Lake Mine, Cameco
Sask. Industry

Much of the gold in Manitoba has been produced as a by-product from base metals mines. The south-east of the province has been especially active near Bissett with reactivation of the old San Antoine Mine. Two other operations have achieved million-ounce status. The Nor-Acme Mine, southwest of Thompson, was found by prospectors in 1938 with the gold in quartz veins and produced 763, 254 tons at 0.142 ounce to the tonne grade from 1949 to 1958. Under new ownership and a name change to New Britannia, the mine was worked from 1994 to 2004 as a joint venture, latterly by Kinross with High River Gold. The mine produced 63,000 ounces in 2003 but was closed the following year due to exhaustion of reserves.

The only current gold mine in Manitoba is located at Bissett in the Rice Lake greenstone belt 256 kilometres north of Winnipeg. The area had twenty-five small mines operating between 1920 and 1939. The San Antonio Mine in the Uchi greenstone belt was discovered in 1911 and produced gold until 1968. There were various attempts at reopening the property with Harmony Gold producing 110,000 ounces prior to 2001. After reorganization and infrastructure work the property reopened under the San Gold banner in mid-2006 with reserves showing 901,800 ounces at a 9.2 gram per tonne grade. Both the Rice Lake and the San Gold #1 mines are promising and 60,000 ounces were produced in 2007.

Generally, Saskatchewan gold miners have worked small properties since Joe Tyrrell noted placer gold on the North Saskatchewan River in 1915; the only notable ones were the Julu and Contact mines. The Seabee Mine, 125 kilometres north of La Ronge, was probed as a development project by Cominco from discovery in 1949 until 1985 when Claude Resources gained control. The mine is now operated as a fly-in operation which extracts ore from narrow veins in volcanic rocks. Since first production in 1991 a total of 718,000 ounces at 6.9 grams per tonne has been recovered and there is a permanent camp site which has undergone both infrastructure development and considerable exploration. Reserves are at 684,000 tonnes and in 2005 44,600 ounces were recovered. The incentive to find more gold is strong. As prospector Eric Partridge said: "You don't have to worry about creating a market for it, like potash. You find it, take it to the mint, and they give you money."

Historically veins have been the single most productive source of gold in British Columbia but presently the metal is more likely to come from association with other commodities in porphyries or VMS deposits. For example, in 2006 only one small mine producing gold from vein structures reopened after 1997 closure due to low gold prices. The Table Mountain Mine of Cusac Gold is in the north of the province and south of Watson Lake. Since first mining in 1979 it has averaged an annual production of 20,420 ounces at a 0.48 grade. The mine is small. Only 315,000 ounces have been given up in its lifetime but it has the advantage of infrastructure in place. Since World War II there have been several small vein gold mines in the province but, although all are now closed, there have been some large producers. The Bridge River camp gave up 4.16 million ounces and had the Bralorne and Pioneer mines as major contributors. The Rossland camp produced 2.7 million ounces with the Center Star and Le Roi mines as the largest properties. The Snip Mine delivered 1.2 million ounces before closure and the Equity Silver Mine, worked primarily for its namesake commodity, refined 508,000 ounces of gold as a by-product. All these properties are being reexamined after many years of closure and the current precious metals prices are spurring dozens of exploration ventures in British Columbia vein properties.

Eskay Creek is located in the mountains, 80 kilometres by air northwest of Stewart, close to the Alaskan Panhandle. The area was first staked by an independent prospector in 1932, and the vicinity was further explored for close to four decades. Further site examination by the Murray Pezim financing group with geologist Chuck Idiziak on the ground revealed a large precious-metal-enriched bonanza. The find proved out to be one of the largest of its kind in the world, a sub-aqueous hot-spring VMS deposit. Homestake Canada bought the property from the Pezim group and took the mine into production from 1971 to 1998, when Homestake was taken over by Barrick Gold. The operation has underground access via three portals using trackless equipment and until recently gold was recovered at a cost of only $49 (U.S.) an ounce. During its first twenty-seven years of operation production was 460,150 tonnes, each averaging 1.43 grams of silver, 29, 6 grams of gold, 0.77% copper and 5.6% zinc. Even with ore reserves predicted to be exhausted by 2008, the mine still delivered 172,000 ounces of gold in 2005. Increased energy costs and difficulty in mining the fringes of the ore body have slowed production since 2006.

The Giant Mine, Yellowknife, N.W.T.
Denis Newman

Yellowknife, in the Dogrib language, is called *Soomba K'e*, or "where the money is," which sums up the location as the mining centre for the Northwest Territories. Historically there have been some large gold mines. Gold was discovered around Yellowknife in 1934, and Cominco did exploration work which resulted of the opening of the Con Mine in 1938. Numerous veins were the commodity source, and the mine worked under three different owners up to 2003. When Miramar shut the property down, the mine had produced 5 million ounces of gold over a 65-year period. The next big mine was found by a prospector in 1935 but not until a massive post-war staking boom took place was the Giant mine established. It produced from 1948 to 2004. The total production was more than 7 million ounces of gold. In its final year it was owned by Miramar. During 1992-3 a bitter strike took place and nine replacement workers were killed by industrial sabotage. The Giant Mine will live on because a mining heritage museum is being built on the site.

The Lupin Mine in Nunavut
Jiri Herman per David General

The Salmita Mine was a small property which operated in the Courageous Lake area. First discovered in 1945, Giant Yellowknife mines worked the property between 1983 and 1987 and produced 179,906 ounces of an unspecified grade. Back in 1960 an Inco exploration party discovered the Lupin deposit near a gossan overlooking Contwoyo Lake, 400 air miles northeast of Yellowknife in the Slave Region. The plant was, for the time, the most northerly gold mine in the world outside of Russia, being only ninety kilometres south of the Arctic Circle. Echo Bay, and then Kinross, operated the Lupin Mine from 1982 to 2005 and were rewarded with 3.37 million ounces of gold over the life of the mine. Wolfden owns the property now and will use the infrastructure in processing other commodities from properties owned in the north. Near Elso, United Keno Hill Mines, owned by Falconbridge, produced high grade silver from several mine sites before closing in 1989 due to depressed silver prices.

Silver

Para desaroller ina mina de oro se necesita un mina de plata.
(It takes a silver mine to make a gold mine.)

Mexican proverb

Silver veins come in three flavours. There are those that have a complex mineral assemblage that includes nickel, cobalt, arsenic and antimony-bearing minerals, as well as silver, but little base metal content: they are known as the Cobalt Suite of Five-Element Veins. The second group has a relatively simple mineral assemblage, with silver, iron, lead and zinc sulphide materials, some barium (barite) and florine (florite) minerals, and are known as the Simple Silver type. Last, there are those that contain very abundant zinc and lead sulphides, and complex silver, silver-arsenic, and silver antimony sulphide minerals, which have the technical name 'clastic metasediment-hosted silver-lead-zinc' type, otherwise known as 'Coeur d'Alene.' A subset of the second type contains uranium as well. All three varieties, except for the uranium-bearing veins in the Great Bear Lake area, occur in early to mid-Proterozoic sedimentary strata, but the veins in all cases are much younger and all formed at the same time as a nearby major intrusive suite. In Canada, only the cobalt type are important to multimine producers, and come from major vein systems in Cobalt and Gowganda, Ontario, the pioneer Silver Islet Mine near Thunder Bay and the Great Bear Lake district in the N.W.T. Minor production came from the veins that belong to the silver simple type near Thunder Bay, Ontario, and significant production came from the Keno Hill mine in the Yukon, a member of the third type.

The cobalt-type veins in all cases are either within a few hundred metres of a major intrusive suite, or cut the intrusions themselves. All veins at Cobalt are both in the Nipissing Diabase and the Huronian sedimentary sequence, and a few extend down into the Archean basement. Individual veins extend for hundreds of metres, and, in addition to the previously mentioned complex sulphide minerals, contain mainly calcite. The Silver Islet vein is similarly a single vein that cuts across a Keweenawan gabbro dyke, one of the suites of intrusions related to the Midcontinental Rift. Only the Great Bear veins sit in a more complex suite of early Proterozoic sedimentary and volcanic rocks, and slightly younger diabase sills. The origin of the fluid that formed these vein systems is poorly understood, but appears to have been through extreme heating of rocks deep in the sedimentary basins, well after those rocks were compacted and deformed. Fluids were generated by dewatering the rocks, a process that occurs on extreme heating, and formed brine capable of transporting the metals. Fluid temperatures may have reached 400 C and they were collected and transported in major rift-related boundary faults. All the vein mineral constituents were probably derived from the sedimentary strata, except uranium, which may have come from the felsic, volcanic and intrusive rocks in the Great Bear area.

Deposition occurred when the fluid pressures exceeded the ability of the faults to remain closed; rapid fracturing and expansion of these fractures within a few hundred metres of the paleosurface may have caused the fluids to boil and the metals to drop out quickly. The simple vein systems probably formed the same way, but at lower temperatures. Mixing with near-surface ground water may have also aided in forming them. Few of these veins are of sufficient size to warrant mining. The clastic metasediment-hosted silver-zinc-lead types are thought to have formed in exactly the same way, from brines formed by high temperature dewatering, transport of major structures, and deposition by boiling.

Early miners who took silver from the ground in the New World of Central and Latin America were not well-informed regarding its strength, malleability, ductility or its electrical and thermal conductivity. They dug it out of the ground mainly for jewellery and decorative purposes. The first prospector who found silver on Skull Rock in 1868 off the shore of Lake Superior, east of what is now Thunder Bay, also recognized that he had found a metal in high demand. The location was a barren rock, now more sympathetically named Silver Islet, which was a mere eighty feet in diameter. Accessed off a peninsula named Sibley after one of the owners of a mining company which developed the prospect, the location was a poor place for mining. The island was battered by storms and what became an underground mine was continually flooded and a miserable place in which to work. But before this first silver mine in Ontario closed in 1884 it gave up $3.5 million in the dollar value of the time, perhaps twenty-five times that in current dollars. The silver from that tiny dot of an island helped move the Lakehead from a fur trading post to a major centre of shipping and commerce.

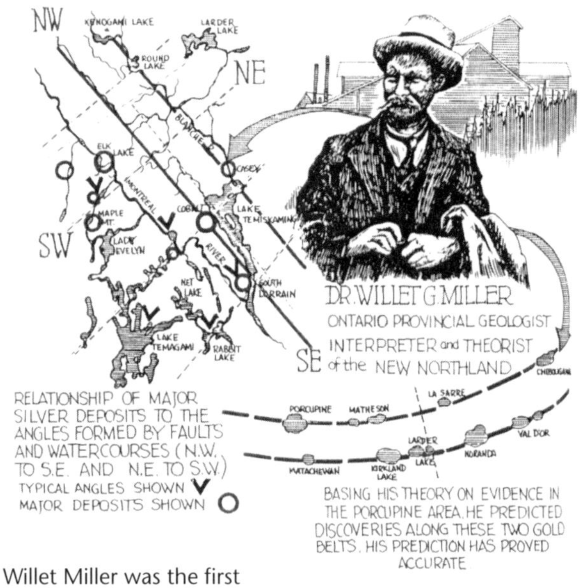

Willet Miller was the first professional geologist on the ground at Cobalt.
Arrow North

The greatest silver discovery in Canada providentially came at a time when it was growing very much in demand for use in photography. The excitement of the Klondike gold workings was fading when the next great metals find in Canada occurred. Silver was discovered in 1903 along the tracks of the provincially-owned Temiskaming and Northern Ontario Railway that was building north from North Bay to the rich clay belt lands surrounding New Liskeard. The line was a colonization effort, meant to bring in settlers and open up Northern Ontario, but the metals discovery by railway workers and prospectors galvanized development of a vast area. Most of the mining in the largely shallow-depth camp was done between 1905 and 1920. The ground was so rich that in 1911, the peak year, 30,000,000 ounces were produced. The town burgeoned in population from a few dozen to between 25,000 to 30,000 souls before dropping down to its present 1,000 residents. In the silver town's heyday, Toronto was said to be the place where you changed trains for Cobalt. The precious metal was taken quickly, often quarried, and after a quarter century of rapid extraction there was practically no mining in the thirties and only a little clean-up production between 1950 and 1960, mainly through the efforts of developers Paul Penna and Lou Cadesky.

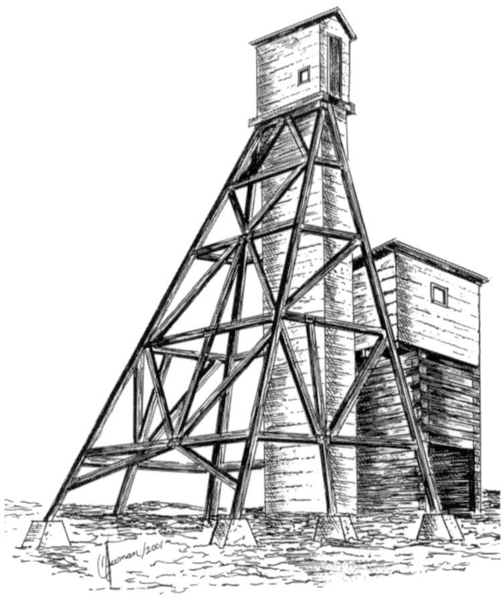

The Nipissing #96 head frame, Cobalt
Denis Newman

Cobalt made the fortunes of scores of individuals, laid the foundation for many mining companies and proved a training ground for miners. Among those who became wealthy was the Timmins family who invested their profits in the Porcupine and the great Hollinger Mine. After establishing that camp, Cobalt money also went on to bankroll much of the Kirkland Lake camp. Cobalt miners worked in smaller silver camps such as Gowganda, where Robert Service spent time long before he went to the Yukon, and South Lorrain. Many technological advances were made in mining and milling in the Cobalt camp but the progress had its down side. The environmental legacy of contaminated soil and water produced at a time when there were no mandated clean-up and closure regulations haunts the area even today. Towns are rarely designated as National Historic Sites, but that honour was accorded to Cobalt in 2002. Today the quiet community has a first rate silver museum and draws tourists to the birthplace of mining in Ontario.

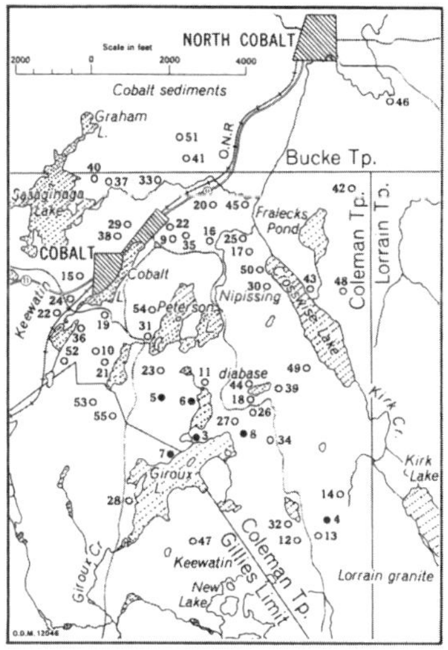

OPERATING MINES ●
(Including those under active development November 1, 1973)

Agnico-Eagle Mines Limited
1. Trout Lake (outside map area)
2. Canadian Keeley (outside map area)
3. Penn Mill
4. Temiskaming

Teck Corporation Limited
5. Silverfields Division

Canadaka Mines Limited
6. Bailey Mill
7. Giroux Lake
8. Conisil

NON-OPERATING MINES ○
(Including dormant properties, past producers, and properties awaiting development)

9. LaRose Mill
10. Nipissing 96 Shaft
11. Penn-Canadian
12. Christopher
13. Cobalt Lode
14. Beaver
15. Buffalo
16. O'Brien
17. Colonial
18. Crown Reserve
19. McKinley-Darragh
20. Nipissing
21. Cart Lake
22. Right-of-Way
23. Nipissing 407
24. Townsite
25. Violet
26. Kerr Lake
27. Lawson
28. Louada (Nu-Silco)
29. Patricia Silver
30. King Edward
31. Trinova (Silver Town)
32. Brady Lake
33. Chambers-Fernland
34. Hargrave
35. LaRose
36. Princess
37. Coleman Township Property
38. Coniagas
39. Drummond
40. Hudson Bay
41. Silver Regent - Genesee
42. Silver Monarch
43. Deer Horn
44. Silver Leaf
45. Mentor
46. Pittsonto Mining Co. Ltd.
47. Professor Silver
48. Rockzone (Smith-Cobalt)
49. Silver Cross
50. Silver Cliff
51. United Cobalt
52. Silver Pack
53. Sudbury Contact
54. Chitaroni Minerals
55. Silver Summit

Outside Map Area
Agaunico
Bomont
Kirkland Townsite
Larum
Millerfields
Langis
Louanna

Mines of the Cobalt Camp.
– Ontario Ministry of Natural Resources

Cobalt mines as of 1973

Silver was discovered in the Great Bear Lake region, 430 kilometres north of Yellowknife, in the Northwest Territories around the time radium was being sought in the late twenties. We remember the area for the uranium found there and used in the Manhattan Project for the first atom bombs. Radium was used in the making of luminous dials for World War II bombers. Three silver mines operated intermittently from 1938 to 1985. These were the Eldorado Mine, Echo Bay Mine, which commenced production in 1964, and the Contact Lake Mine.

The O'Brien Mine, 1904-1968, Cobalt
Denis Newman

Mine finder Bob Jowsey and friends *PDAC*

Many Cobalt mines were not of architectural merit. *PDAC*

Lunch for the exploration crew at Hackett River *Sabina Silver*

Silver is found elsewhere in this book for it associates itself with gold and base metals. We often think of the metal as a by-product, since, out of the more than one billion kilograms of silver mined annually in Canada, the vast majority comes along as a happy sideline, a bonus from other mining endeavours. For a long time Eskay Creek in British Columbia has mined the most silver, delivering roughly 28% of the total produced in the country. Sabina Silver is a junior company with its sights set on mining silver as its main interest. Its claims at Hackett River, south of Bathurst Inlet in Nunavut, as outlined in a 2006 NI43-101 report, indicated resources of 47 million tonnes of ore containing 205 million ounces of silver, as well as 4.3 billion pounds of zinc, 330 million pounds of copper, 664 million pounds of lead and 433,000 ounces of gold. The deposit is in the Hackett River greenstone belt in massive sulphides. The ground has been explored over half a century by three previous interests, Teck Cominco being the last before the present operator took the ground. To date, the design calls for two open pits and one decline underground with construction to be concluded by 2011. Mining company Silver Wheaton is a major shareholder in Sabina. The project is dependent on a 211 kilometre all-weather access road being built to Bathurst Inlet, which is supported by three other mining companies and the Inuit community.

What Goes Around

Economic Cycles and Exploration Budgets

Exploration is a creative act and no matter how much is said about the potential natural resources of a country, they are worth nothing until they have been discovered.

Dr. Stanley Holmes, scientific entrepreneur

Mining is ever plagued by external, worldwide influences. Recessions as seen in the 1890s and the Great Depression of the 'dirty thirties' are ready examples. Industry observers recall the booms of the 1950s, 1970s and 1990s, and the busts that deflated them in the 1960s, 1980s and the early part of the present century. Charles Graham of CAMIRO put it this way: "We are continually trapped by the commodity price cycle. When prices are up, as today, the industry has no time to innovate and when prices are down, there is no money." He was talking about mining technology development but the commodity cycle affects all facets of the industry.

Worldwide economic slowdowns reduce demand for metals and consequently depress their prices. Third World countries are increasing their stake in the mining business and developing their resources, actually flooding the commodity market. Ironically much of this development comes through financial support from Western agencies. The advantages these overseas producers enjoy are lower wages, often a more favourable climate, and open pit extraction, which is usually less costly to operate than Canadian underground facilities. During the bust time of the early eighties, Ontario's metal production dropped from $3.8 billion to $2.7 billion. This, plus a loss of 6,000 mining jobs over a two-year period, caused the national economy to take a significant hit. The

Schumacher honours its founders with monuments. The great McIntyre Mine can be seen in the background.
Brenda Burey

pundits fill up financial columns now and again with referrences to mining as a 'sunset industry.' Whenever that sort of depressing news comes to the fore, there is usually a counterbalance. For example, it was in the mid-eighties that the Hemlo gold mines came on stream.

Anyonewhohasworkedinorlivedaround the mining industry has seen first hand the results of a boom and bust economy in small communities such as Val d'Or, Quebec. In cities, those with skills, an honest outlook and a strong work ethic can usually make a good living. These attributes are also valued by people in small frontier towns across Canada, such as Val d'Or, but mining and lumber communities are directly affected by distant world markets.

Downturns in demand can decimate small business and reduce employment in resource-dependent towns, making civic services difficult to maintain and resulting in an exodus of citizens in search of badly needed work. In 1996 at the height of the last gold boom, the town named for the Valley of Gold ranked first in Quebec in terms of per capita sector investment. Three years later declining gold prices saw the exit from the community of two major gold producers and some small mines closed. Young adults left town, the population shrank four percent and income levels dropped. Then gold started an upward climb and boom times came again to the northwestern Quebec town. Similar stories may be found across the north country.

The Metal Economics Group, based in Halifax, Nova Scotia, is a world leader in mining industry intelligence. The private think tank, funded by industry, banking interests and government, is a primary source of information and analysis on global mining exploration, development and production, strategic planning and acquisitions. When the MEG speaks, key mining industry figures listen. A report by the group in mid 2006 concluded that the decline in gold reserves discovered through exploration could result in shortages in the future. The good news in the report is that the world's twenty largest producers have used acquisition and exploration to maintain a production life of fourteen years in reserves at 2005 production rates. But declining gold replacement since 1997 via exploration, coupled with a drop in exploration spending, could result in gold scarcity in future. The majors consider 2.5 million ounces as a large gold deposit worth developing, but not enough of these have been found to maintain major sources for the metal.

Metals Economics Group news is not always dour. In 2003 its press release acknowledged that worldwide exploration spending had risen for the first time in six years. Spending on exploration had risen from a ten-year low of $1.9 billion in 2002 to $2.9 billion, a figure obtained by analyzing 917 companies' exploration budgets. Cutbacks by the majors, negative industry consolidation effects and loss of funding for several junior companies all took their toll. The MEG concluded that gold price buoyancy and strong demand for base metals should continue to support rising exploration budgets. An interesting side note was that Australian organizations were lobbying their national government to adopt some form of flow-through share program similar to Canada's. The in-depth analysis and projections produced by the Halifax-based group are costly — way beyond the means of all but industry stakeholders with extremely deep pockets. A recent 647-page, two volume study by the MEG on Corporate Exploration Strategies cost $11,000 retail.

Minerals and Mines of the Yukon

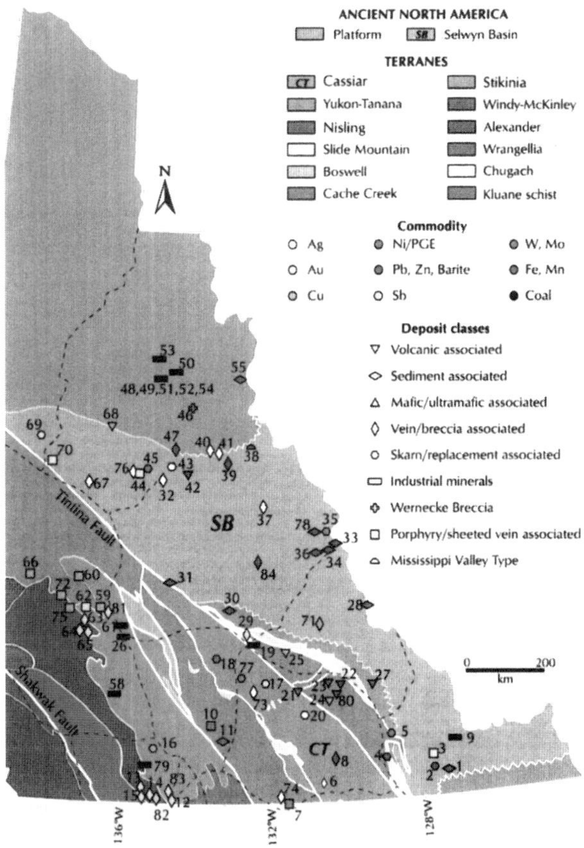

The scramble for uranium is one example of a commodity for which demand outstrips production and accelerates its own cycle. The market for yellowcake is projected to increase at an annual rate of 2.8% to 2010 and double by 2020. Currently commercial nuclear power plants produce 16% of the world's energy requirements. These plants require

178 million pounds of U_3O_8 per year, and within the next ten years it is expected that, with new facilities coming on stream, the demand for uranium will increase by one-third. To meet this demand the mining industry must find deposits and increase output by 60%. Exploration activity for the commodity in Canada rivals diamonds in scale. The fact that uranium's price per pound has skyrocketed has been an additional spur to activity.

Analyst Jim Jubak has put forward the proposition that the boom-bust cycle has been put aside at present and that the next bust has been delayed for a long time. He feels global economic growth will accelerate through 2008 and that the momentum will avoid the bust that could occur the following year.

Charging Bull: a symbol of aggressive financial optimism that sits in Bowling Green Park near Wall Street in New York City.
Arturo Di Modica

According to popular experience, mining companies go through a period of progressively lower spending on exploration coupled with decline in production. Then there is a demand surge for metals and industry players struggle to increase production and beef up their exploration teams. This optimistic period is followed by eventual over production and a flattening of commodity prices. Certainly a huge cash flow has resulted this time around. Citigroup economists project that mining companies will generate $253 billion in cash from 2006 through the end of 2008. But many major companies have not increased production. A glance at BHP Billiton news releases indicates that while the mining giant had an operating cash flow of $10.8 billion at the end of fiscal 2006, its projected capital spending was estimated to be half that figure.

The City of Sudbury seized the initiative when Xstrata and CVRD took over mining in the basin. It welcomed the newcomers in a commissioned paper called *Claiming our Stake! Building a Sustainable Community*. In this document the nickel capital also postulated the Commodity Super Cycle. This effect was described as an extended decades-long trend rise in commodity prices driven by a rise in world urbanization and industrialization. The paper quotes the Australian Institute of Mining and Metallurgy as being of the opinion that over the next half century the world will use five times the mineral resources that have been mined to the year 2000. The key to this super cycle is the growth of demand for base metals from China and other Asian countries. For nickel producers such as Sudbury, the growing middle class markets of the East with their rapidly increasing demand for metals present both challenges and opportunities.

In 2007 the Metals Economics Group took another look at Canadian exploration strategies and reported that exploration budgets have increased for the fourth year in a row to more than $7.1 billion, the highest level since studies of exploration strategies began in 1989. This was based on an assessment of 1,600 companies. Base metals and silver exploration budgets have increased substantially over gold. Late stage exploration spending has been accelerated as companies push to bring projects to a production decision.

No one denies the existence of economic cycles and their impact upon exploration budgets. The current debate could be about the possibility that there is indeed a mining supercycle. Examining the country's mining industry at this time, Canada is the world's largest exporter of metals; the leading supplier of uranium; the third largest in the world for production of platinum-group metals, titanium and zinc; fifth in copper, lead, silver and gold; and a leading producer of nickel. There is huge growth potential, as evidenced by the vast nickel, copper and cobalt deposit at Voisey's Bay and by the growth in diamond

The Cusac Gold mine is in the Cassiar camp in north central B.C.
Cusac G.M.

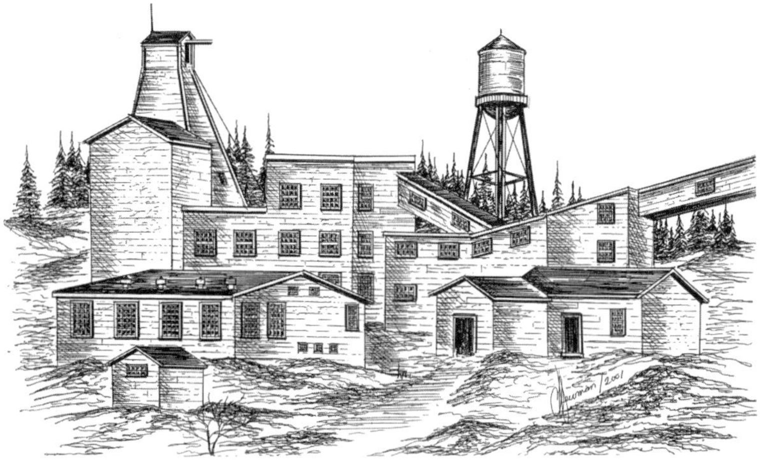

The Barvue Mine, QC, 1953
Denis Newman

production. Metals production is primarily export-oriented and as much as 90% of some commodities go to external markets. Minerals, metals and energy sectors contribute about 9% of the Gross Domestic Product. Investor confidence in the mining sector is high, in large part due to the standards of disclosure for mineral projects. There are 3,000 domestic and 150 foreign companies working in the industry. One tenth of these are actively mining, while the rest are involved in exploration, development or raising funds for future work. The industry plays a huge role in boosting Canada's economy.

Economic cycles generated by circumstances beyond the control of the mining industry have always been cause for concern. One way of combating the spectre of boom and bust uncertainty is to examine background data which should carry the mining business forward for the near future. Canada's aggregate metal production has not declined significantly over the past thirty years. Nonetheless, there are specific geographic areas of concern for metal mining. They are in Bathurst, New Brunswick, the Abitibi region of Quebec, British Columbia, the Yukon and Nunavut. The massive sulphide deposits in New Brunswick are running out, and the area requires intense deep exploration to garner new prospects. The Abitibi belt of Quebec has been extensively mined and reserves are slipping, but several gold mines have been reopened with the advent of high prices and, while exploration is ongoing, intense activity is taking place in the James Bay area. The problem with west coast porphyry deposits is that the copper, molybdenum and gold grades are too low to compete with huge South American ventures in places such as Chile. Exploration in British Columbia is trending further north in search of higher grade material. Both Nunavut and the Yukon have seen the closure of key mines, and much attention has been diverted to promising discoveries in recent years, although lack of infrastructure has always been a deterrent to mining investment in those remote northern regions.

The Nautilus drill ship is one mining initiative for the future.

Nautilus Minerals

The old adage that one has to spend a dollar to make a dollar is a prime guide for exploration budget makers in Canada. Fortunately Canadian miners are optimists. Core values in the industry are still ideas, imagination and initiative. The real strength of the Canadian exploration system has been its diversity. New mineral wealth is sought in many places using different methods and approaches because each potential ore body has its own characteristics. Ways of reading the rocks are as varied as the geologists, prospectors and exploration firms that venture across this vast country. Those in exploration cannot be shackled by economic cycles. Exploration activity is highly innovative and continues to require creativity and a determination to succeed, qualities that have never been mined-out in Canada.

Mining Leads the Way

Some Summary Thoughts

Don't gamble; take all your savings and buy some good stock and hold it till it goes up, then sell it. If it don't go up, don't buy it.

Will Rogers (1879 – 1935)

This wry financial advice was offered by the great American humorist towards the end of his life, so close to the economic devastation rendered by the Great Depression. Thankfully no important decisions in investment today are ever quite that contradictory.

Let us start by reviewing facts about the Canadian mining industry with which everyone will agree. Mining is a basic industry which is vital to the world's economy. Most of the commodities provide the basis for the fabrication of items for everyday use. The energy sources we require come from the earth. Since the mineral wealth obtained by mining on our planet are vital to our health, security and well-being, a strong and technologically advanced, well-managed mining industry is essential to our future.

There is much to be positive about in Canadian mining; the country is the largest exporter of metals and minerals in the world. Canada is home to sixty percent of all the world's exploration companies. Financial institutions in this country raise half of all equity financing for mining projects. As for commodities, Canadian diamonds are ranked fourth globally by value of production and, of the sixty commodities Canada produces, seventeen rank in the top ten in production worldwide. The industry is safe and the country has a stable political environment. In its report,

Doing Business in 2007, the World Bank ranks Canada an impressive fourth out of 175 countries in regulatory environments. We have free movement of goods, services and information. There is a stable economy and banking system. Toronto is acknowledged as the mining investment capital of the world and Vancouver is not far behind. Investment is not placed in jeopardy due to the capricious whims of newly-elected governments with political agendas directed against free enterprise. In what *The Northern Miner* calls "reddening the risk map," countries such as Russia, Mongolia, China, several Latin American states and other authoritarian regimes, have all provided unpleasant surprises for foreigners trying to do business in the mining field. Political stabilization and mining-friendly jurisdictions have opened new business doors for Canada. Environmentally, Canadian standards are among the highest on the planet today and there are many regulations in areas such as safety and mine reclamation that make the country a world leader in humane and responsible resource environments.

New gold diggings in Tangier, Nova Scotia, 1861

Nova Scotia Archives

News stories about the exploitation of Aboriginal people around the world with regard to the use of natural resources appear frequently in the media. In Canada, both government and private industry acknowledge First Nations peoples as primary stakeholders in the land, and not only are they involved in resource development at all stages, but they are also actively encouraged to become active stakeholders in mining enterprises. Nowhere is this clearer than in the landmark publication *Mining Information Kit* published jointly by the Government of Canada, the PDAC, the Mining Association of Canada and the Canadian Aboriginal Minerals Association. Many First Nations groups are in highly profitable businesses with the mining companies they host, providing employees, air transport services, camp facilities and catering.

The Inuit have been quick to seize economic opportunities in this regard. Generally companies have developed good relations with Aboriginal people but there must be clearly understood and enforced guidelines for land use. The recent situation in northwestern Ontario involving a junior miner is a case in point where the province permitted exploration and the local band obstructed and closed down an exploration camp.

The focus on environmental protection is strong. The protocol ***e3***, for Environmental Excellence in Exploration, originated in Canada and is the world leader in the field of resource environmental protection. An initiative of major mining companies, and managed and directed by the Prospectors and Developers Association of Canada, the influence of this endeavour is growing in leaps and bounds. ***E3*** outlines environmental stewardship in the exploration stage of mineral development by outlining sound environmental management practices which are being applied in a growing number of countries throughout the world.

Mineral discoveries result in new planned towns like Manitouwadge, ON.

Mining communities stand out in promoting and reinforcing the industry that provides their bread and butter. The Canadian nickel capital of Sudbury is just one example. The city understands that it is sitting on a highly valued resource and boasts of the only research university hosted by a mining community, Laurentian University. Another industry-specific facility found in the nickel belt is the Northern Centre for Advanced Technology. The newly instituted MERC — Mineral Exploration Research Centre — is also based in the nickel city. In a bid to attract more federal funding, the City of Sudbury has commissioned a wide variety of initiatives showcasing advantages the region has, such as mining research and innovation for the metals industry. As a result, the city is well on the way to promoting itself as a global centre of hard-rock mining excellence. In other locations private industry has backed mining education. The former International Nickel Company founded a mining innovation centre at Memorial University, Newfoundland. A federal government partnership enabled development of the Aluminium Technology Centre on the Lac-St. Jean-Saguenay campus of the University of Quebec.

Think tanks can make dubious predictions. The prestigious Club of Rome must have been drinking too much grappa at lunch one day in 1972. The group pronounced in its paper *The Limits to Growth* that at exponential growth rates the world would run out of gold by 1981, tin by 1987, zinc by 1990, petroleum by 1992, and copper, lead and natural gas in 1993. These conclusions sound ridiculous now with the benefit of hindsight, but there is in reality a crisis in the delivery of commodities looming on the not-too-distant horizon.

An exploration crew hams it up in Ketchikan, Alaska, en route to Galore Creek, 1964. Left to right: Hugh Stewart, Ross Blusson, Ted Carpenter and Harold Cuddie.
David Barr

Another potential source for disruption of the mining industry is the well-organized coalition of environmental groups that ignore their own needs for metals and are extremely vocal in opposing the industry in general and new mining in particular. The arguments they use are often based on past experience and carefully ignore the developments in environmental protection and reclamation that have taken place over the past sixty years. The Pembina Institute published an assessment of the value of public support of the metal mining industry in Canada in 2002. The paper examined the level of government funds in strengthening the mining industry. Quebec is singled out for the extent of its equity investments in the junior exploration companies. The report concludes that flow-through shares should be terminated; there should be no tax relief for the industry; and governments should not provide infrastructure to facilitate the conduct of mining operations. The paper suggests that public resources would be better employed in promoting forms of economic endeavour which are less vulnerable to shifts in economic conditions. This polemic published under the guise of a balanced study ignores the fact that Canada must be competitive in mining as in any other industry and that, even with extensive recycling, there will still be demand for metals. Maybe these people would not require nickel for their stainless steel sinks as they revert to the horse and buggy days, but recall that even in those times folks used iron. The mining industry in Canada must be vigilant and combat such attacks on its work with a vigorous rebuttal of one-sided advocacy groups.

There is a severe reserve problem facing the Canadian mining industry. The Mining Association of Canada, in a 2007 submission to the federal Finance Committee which is responsible for the national budget, stated that levels of proven and probable reserves in key minerals and metals have declined by 50 to 80 percent since 1980. The Metals Economics Group warns that world nonferrous exploration budgets fell to a twelve-year low in 2002 and, although the hunt for minerals is vigorous at present, the exploration business still fluctuates widely, relative to the mining cycle. In recent years imports of metal concentrates to produce finished metal products have exceeded exports.

Students at Northwest Community College, Smithers, B.C. on an exploration studies field trip
NWC College

Canadians must applaud and support federal initiatives to strengthen our claims to Arctic sovereignty. Great mineral deposits of the future will be found in less-travelled places, mountains and the far north. The economic and strategic value of northern resource development is critical to the continued development of the nation. Industry well-wishers will no doubt remind our legislators that although the Mining Exploration Tax Credit or the 'super' flow-through program has been extended, the much-needed initiative of the commitment to the Cooperative Geological Mapping Strategies has been on the books since 2000 and has not been honoured by government. Canada has ten million square kilometres of land area and probably a half of that has mineral potential which may be discovered if geological mapping and other geoscientific initiatives are funded on a consistent basis.

The mining industry is dependent on huge investment for production. Moves that ensure stability are therefore essential. Government funding and participation in various areas of the mining cycle are necessary to ensure public and investor confidence. Canada presently has no national securities regulatory body that would simplify and help attract capital investment. There have also been calls for a research and development coordinator to ensure that there is a consistent approach to solving problems as they arise and raise the alarm when political commitments are not fulfilled. The holder of that office might examine the uneven exploration patterns where gold, diamonds and uranium are at the top of the heap and vital base metals exploration get relatively short shrift. Tax incentives such as a suspension or rebate of corporate taxes in the early life of a mine have been in place in the past and certainly the Mattabi Mine in Ontario and others achieved production in 1972 due to this significant government incentive.

The mining resource sector is structured around senior and junior mining companies. Almost half of exploration funds are expended by juniors who gamble more readily and hustle to locate promising deposits. Seniors have the funds but are much more deliberate and take less risk in their approach to investment. Flow-through funding is of great benefit to smaller companies and the tax benefits attract investment, but the subsidy, as it now operates, encourages investors to reap the tax benefits and then dump stock. There is less opportunity offered by government for senior firms to write off exploration expenditures. One suggestion to remedy this inequity is the provision of tax holidays to fund the huge cost of mine development. Another is to examine the Canadian Exploration Expense category of the Income Tax Act to encourage deep exploration in existing confirmed deposit areas. Too often firms go after remnants of deposits in previously mined-out properties to raise cash while avoiding the enormous cost of new exploration and development. Senior companies should be consistent in maintaining strong, successful exploration departments because more cash-strapped juniors often only employ geologists, engineers and other scientists when required for specific projects in hand. In the past the large exploration departments maintained by the seniors provided opportunities for summer student employment which promoted a strong, professional labour force for Canada's mines of the future.

The Dome Mine, South Porcupine, began producing in 1910. This #3 shaft worked from 1920-1985.

Denis Newman

A trend which has some worrisome implications for a healthy mineral exploration climate is the growth of major acquisitions by companies such as Goldcorp, Barrick, Iamgold, Kinross, Xstrata and CVRD. The latter two mining giants are foreign-owned. As historic miners such as Inco and Falconbridge are swallowed up, national control of a great strategic industry is weakened. There is clear evidence that such takeovers, Canadian-inspired or international, lead to a decline in exploration expenditures of the firms involved, again jeopardizing our future mining labour supply as less casual student labour is required in the North.

Eira Thomas, seen with her brother at the site of a kimberlite discovery in the mid 1990s, looks quite different from her magazine cover portrait.
Stornoway Diamonds

A recent editorial in the *Canadian Mining Journal* rightly places the blame for the current and upcoming shortage of trained workers, technologists and engineers on the mining industry. Government should also bear some of the responsibility for the decline in the prospective employee pool. Between 1995 and 2002 the output of mining engineering graduates from Canadian universities was only 109. It is predicted that all levels of the industry will need 81,000 new technical and professional employees to replace anticipated retirements, according to the Mining in Transition and Adjustment Council. Prospective science students will migrate to other studies leading to secure employment unless they are warmly recruited. Fortunately, several education institutions are making headway in this area. Community colleges such as Northwest Community College at Smithers, B.C., and Ontario's Haileybury School of Mines, are beefing up their course offerings, and many of these are aimed at First Nations students in disciplines such as underground technician, surveying, diamond drilling, prospecting and reclamation programs. The Department of Geology at the University of Toronto has launched The Explorers Fund, aimed at securing donations to offer future geoscientists support for their studies. The Mining Industry Human Resources Council has secured federal government support for a mining labour market transition project which will assist in drawing workers to the mining industry from sectors with an excess supply of labour. Charitable organizations such as Mining Matters, a branch of the PDAC, can play a larger role in educating students in the sciences about the opportunities in the mining industry. The new federal government's Foreign Credentials Referral Office (FCRO) should make it easier for international mining professionals to work in Canada.

The mining industry needs to focus more on public relations to ensure a fair and balanced public discussion of resource requirements and problems. Attempts to obtain permits and other processes leading to resource extraction are often thwarted due to a number of factors, not the least of which are political interference, environmentalists and various stripes of activism. One example in recent memory is the giant Windy Craggy VMS copper deposit in the Cassiar region in the northern part of British Columbia. The mining company involved was well aware that the area was a sensitive environment but its efforts to explain how the location would be protected were drowned in a sea of protests. The result was that the provincial government withdrew the area from further exploration and in 1993 turned it into a provincial park. Those who fight resource development are well-organized, employing buzz phrases such as "Dirty Metals" and "Life before Profit." They are highly vocal and have funds provided by a large number of environmental and other groups. They totally ignore the ***e3*** environmental protocols and the attempts of mining companies to carry out progressive social policies. Long before there were any environmental rules, the late Dr. Norman Keevil Sr. discovered an extremely rich copper deposit on an island in the middle of the beautiful Northern Ontario district of Temagami. Despite having practically no legislation to hinder mining, the deposit was mined for close to twenty years and then the site was restored in such a way that visitors today cannot even see where the ground was disturbed. Mining associations and companies must become more proficient and aggressive in the business of explaining their work and telling the public relations story.

Despite the naysayers who are hostile to the industry, Canadian miners do a good job of providing economic metals, uranium and diamonds which enhance the lives of fellow citizens and contribute hugely to the national economy. This book is a modest attempt to shed some light on the economic successes resulting from successful exploration programs and technological advances in the mining industry since World War II. We have reviewed developments in the geosciences which aid exploration

and ore body detection and have examined some of the mineral discoveries resulting from the work of geologists and prospectors in the search for fortunes in the ground. Right now business is on the upswing. There is a huge overseas demand, particularly from Asia, for the products of our mines. In the foreseeable future there does not seem to be any downturn in the market. The paper *Facts and Figures 2006* published by the Mining Association of Canada underlines the importance of mining to the nation's economy. The industry employs 388,000 people, contributes $42 billion to the gross national domestic product and accounts for 14% of Canadian exports. To maintain this output the mining industry must replenish reserves, make new discoveries and commit in every way possible to investment in mineral exploration in Canada. Mining is a non-stop activity. Twenty-four hours a day there are people exploring, mining, milling or planning for mining in Canada. The country's greatest asset is its natural resources awaiting development. And for our small population density in most areas, particularly of the North, there is no other base upon which to develop and maintain a healthy and sustainable living for its people.

We conclude with the words of respect for the exploration business written by the man generally acknowledged to be Canada's greatest scientist, Sir William Logan, 1799-1875: "I have dined with lords and ladies, chatted with Queen Victoria, and been formally received by the Emperor Napoleon, yet my most cherished memories come not from the fine salons of Europe, but from a leaky tent, a bark canoe, my rock hammer, compass and theodolite, and the vast and mysterious wilderness of Canada."

Mines of Canada: 1946 -2006

MINE NAME	LOCATION	PRODUCER	MINE LIFE SPAN	PRODUCTION

FAULTS AND FISSURES (GOLD)

NEWFOUNDLAND

Hope Brook	Chetwynd	BP Selco-Royal Oak	1987-1991, 1992-1997	752,162 oz. Au total produced
Stogertite	Springdale		1996	1,952 oz. Au total produced
Nugget Pond	Baie Verte	Richmont	1997-1999	123,793 oz Au total produced
Hammerdown	Kings Pt	Richmont	2001-2004	143,000 oz. Au total produced
Beaver Brook	Glenwood	Roycefield	1997-1998	23,741 tonnes @ 3.91% Sb
Renfrew		Horne Mining		220,000 tons @ 0.28 opt Au
Forest Hill		Seabright		93,333 tons @ 0.38 opt Au
Caribou		Cominco		290,000 tons @ 0.30 opt Au
Tangier		Coxheath		124,000 tons @ 0.28 opt Au
Cochrane Hill		Northumberland		9,150 tons @ 0.22 opt Au

QUEBEC

CHIBOUGAMAU CAMP				
Joe Mann	Rohault Twp.	Key Anacon-Campbell Resources	1956-59, 74-75, 2002-Present	745,000 tonnes @ 3.60 gpt Au to 1975
Norbeau	McKenzie Twp.	Norbeau Mines		380,000 tonnes @ 13.98 gpt Au, 1.88 gpt Ag
Lac Gwillim	McKenzie Twp.	Campbell Res.		228,000 tonnes @ 4.81 gpt Au, 5.08 gpt Ag
Bachelor Lake	Lasueur Twp.	Bachelor Lake Mines	1982-1988	131,000 oz
Lac Short	Desmaraisvile		1983-1992	
Montauban	Montauban Twp.	Muscocho Mines	1983-1990	
Troilus		Placer-Inmet	1997-Present	
Belleterre	Guillet Twp.	Belleterre Que. Mines	1936-1958	2.4 mil tonnes @ 10.73 gpt Au, 1.37 gpt Ag
JOUTEL CAMP				
Eagle	Joutel	Agnico Eagle	1974-	
Telbel	Joutel	Agnico Eagle	1976-1993	
Sleeping Giant	Joutel	Aurizon-Cambior-IAM Gold	1987-Present	
Golden Pond	Casa Beradi	Inco-Aurizon	1989-1992, 2006-Present	3.45 mil. tonnes @ 0.20 gpt Au
VAL D'OR CAMP				
Quesaba			1949-	
Siscoe	Dubuisson Twp.	Siscoe Gold	1929-1949	3.0 mil. tonnes @ 9.81 gpt Au, 0.75 gpt Ag
Simkar			1947-1949, 1991-	
Perron	Pascalis Twp.	Perron Gold Mines	1933-1951	1.6 mil. tonnes @ 8.64 gpt Au
Beaufor		Louvem, Richmont	1933, 1956, 1996-Present	266,793 tonns @ 6.22gpt Au
Bevcourt			1952-	
Kiena	Dubuisson Twp.	Placer-Wesdome	1981-2002	1.4 mil. tonnes @ 4.50 gpt Au
Dumont	Bourlamaque Twp.	Aur Res.	1980-1994	1.2 mil. tonnes @ 6.51 gpt Au
Balmoral	Bourlamaque Twp.	Belmoral	1981-1994	1.6 mil. tonnes @ 6.99 gpt Au
Sigma	Bourlamaque Twp.	Sigma-McWatters-Century	1937-2001, 2005-Present	24.8 mil. tonnes @ 5.48 gpt Au to 2001
Lamacque	Bourlamaque Twp.	Lamacque-(Teck Corp)	1935-1985	24.2 mil. tonnes @ 5.81 gpt Au
Beacon #2	Louvicourt Twp.	Dorval	1980-1981, 1986-1996	132,328 tons @ 2.41 gpt Au
L.C.Belliveau	Louvicourt Twp.		1989-1992	
Chimo	Vauquelin Twp.	Chimo Gold-Louvem Cambior	1966-1967, 1988-1996	2.4 mil. tonnes @4.7 gpt Au
Norlartic	Vassan Twp.		1959-1963, 1991-1993	1.5 mil. tonnes @ 4.23 gpt Au

MINE NAME	LOCATION	PRODUCER	MINE LIFE SPAN	PRODUCTION
Croinor	Pershing Twp.	South Malartic	1986-1997, 2004-Present	
Joubi	Dubuisson Twp.		1989-1991	
Sullivan	Dubuisson Twp.	Sullivan Mines	1934-1967	4.6 mil. tonnes @ 7.92 gpt Au, 2.40 gpt Ag
Bevcon-Buffadison	Louvicourt Twp.	Bevcon Mines	1947-1965	3.0 mll.tonnes @ 4.63 gpt Au, 1.95 gpt Ag
Akasaba	Louvicourt Twp.	Soquem	1960-1963	289,428 tonnes @ 5.18 gpt Au, 1.47 gpt Ag
Wrightbar	Louvicourt Twp.	Que-Manitou	1986-1988, 1995-1996	304,657 tonnes @ 7.8 gpt Au
Little Long Lac	Dubuisson Twp.		1964-1969	25, 000 tonnes @ 8.00 gpt Au
Marban	Dubuisson Twp.	Malartic Goldfields	1961-1974	2.0 mil. tonnes @ 5.40 gpt Au, 0.62 gpt Ag
Goldex	Dubuisson Twp.		1973-1975, 1980	
Orenada	Bourlamaque Twp.	Mines d'Or et Cuivre Mid Canada	1980-1981, 1994	
Bevcon	Louvicourt Twp.			
Goldstack	Dubuisson Twp.		1935-1965	
Louvicourt Goldfields	Louvicourt Twp.	Louvicourt Gold Mines	1947-1949, 1990-1992	
Bras d'Or	Bourlamaque Twp.	Belmoral Gold Mines	1979-	
MALARTIC CAMP				
Malartic Gold	Fourniere Twp.	Malartic Goldfields	1935-1965	9.0 mil. tonnes @ 6.17 gpt Au, 0.31 gpt Ag
Camflo	Malartic Twp.	Camflo-Barrick	1981-1992	
Malartic Hygrade	Malartic Twp.	Malartic Gold	-1992	
Canadian Malartic	Fourniere Twp.	Lac Minerals	1935-1965	9.9 mil. tonnes @ 3.77 gpt Au, 2.47 gpt Ag
Kierns	Malartic Twp.	Aur Res	1965-1966, 1989-1992	251,015 tonnes @ 6.30 gpt Au
East Amphi	Malartic Twp.	Richmont	1998-1999	
Barnat-Sladen	Fourniere Twp.	Barnat Gold	1938-1970, 1979-1981	8.4 mil. tonnes @ 4.73 gpt Au, 1.17 gpt Ag
East Malartic	Fourniere Twp.	Lac Minerals	1938-1985	18.0 mil. tonnes @ 5.19 gpt Au, 1.27 gpt Ag
Orion			1988-1990	
CADILLAC CAMP				
Wood Cadillac	Cadillac Twp.		1939-1942, 1947-1949	814,970 tonnes @ 4.8 gpt Au
Darius	Cadillac Twp.	O'Brien	1933-1956, 1978-1981	1.27 mil. tonnes @ 16.1 gpt Au
Mouska	Bousquet Twp.	Lac Minerals-Cambior	1942-1941, 1980-Present	1.16 mil. tonnes @ 10.9 gpt Au, 1.7 gpt Ag
Bousquet #1	Bousquet Twp.	Lac Minerals	1979-1996	7.5 mil tonnes @ 5.3 gpt Au
Bousquet #2	Bousquet Twp.	Lac Minerals	1990-2002	8.1 mil tonnes @ 8.3 gpt Au, 0.57% Cu
Thomson-Bousquet	Bousquet Twp.	Lac Minerals	1979-	
Mic Mac	Bousquet Twp.	Mic Mac Mines	1942-1949	0.7 mil. tonnes @ 4.6 gpt Au, 0.11 gpt Ag
Doyon	Bousquet Twp.	Lac Minerals-Cambior	1980-	26.8 mil. tonnes @ 5.8 gpt Au to 2005
Lapa	Cadillac Twp.		1939-1949	346,000 tonnes @ 4.3 gpt Au
Dumagami	Bousquet Twp.		1988-2000	6.5 mil tonnes @ 6.8 gpt Au
NORANDA CAMP				
Francoeur	Beauchastel Twp.	Wright-Hargreaves-Richmont	1938-47, 1968-71, 91-2001	1.0 mil.tonnes @ 6.17 gpt Au to 1971
Donalda	Rouyn Twp.	Kerralda Mines	1948-55, 70-71, 93-94	732,tonnes @ 6.17 gpt Au, 6.22 gpt Ag
Anglo-Rouyn	Rouyn Twp.		1949-1951	132,000 tonnes @ 8.60 gpt Au, 4.73 gpt Au
Powell-Rouyn	Rouyn Twp.	Lac Minerals	1937-1956	2.7 mil. tonnes @ 4.43 gpt Au
Chadbourne	Rouyn Twp.	Noranda	1979-1985	
Peel-Elder	Beauchastel Twp.		1946-1966, 1988-1989	
P.Beachemin	Duprat Twp.		1955-62, 1989-92	
Silidor	Rouyn Twp.	Noranda-Battle Mountain	1990-1997	
Granada			1930-35, 1993-94	
Astoria			1992-1955	
Hosco	Joannes Twp.	Hosco Gold	1948-1949	46,000 tonnes @ 5.69 gpt Au
Rouyn-Merger	Rouyn Twp.	Soquem	1948-1949	29,000 tonnes @ 4.11 gpt Au
Senator	Rouyn Twp.	New Senator Rouyn	1940-1955	1.8 mil. tonnes @ 4.70 gpt Au, 1.13 gpt Ag
Stadacona	Rouyn Twp.	Falconbridge	1936-1958	3.0 mil. tonnes @ 5.25 gpt Au, 1.37 gpt Ag
Wasa Lake		Lac Minerals	1965-1971	2.1 mil. tonnes @ 4.29 gpt Au
Don Rouyn	Rouyn Twp.	Noranda	1957-1980	5.2 mil. tonnes @ 0.15% Cu, (Flux)
Elder	Beachestel Twp.	Aunore Res.	1944-1965	1.9 mil. tonnes @ 5.31 gpt Au
Eidrich-Flavel	Duprat Twp.	Sullivan Mines	1956-1962	651,000 tonnes @ 4.94 gpt Au
New Marlon	Rouyn Twp.	Noranda	1947-1949	98,000 tonnes @ 6.14 gpt Au

MINE NAME	LOCATION	PRODUCER	MINE LIFE SPAN	PRODUCTION
Quesabe	Duprat Twp.	Sunburst Expl.	1949-1952	89,000 tonnes @ 10.72 gpt Au, 4.42 gpt Ag
Beattie Mine	Duparquet Twp.	Beattie-Duquesne	1937-1956	9.6 mil. tonnes @ 4.02 gpt Au, 0.99 gpt Ag
El Coco	Dasseret Twp.		1981-1982	29,000 tonnes @ 2.51 gpt Au
Montbray	Rouyn Twp.		1989-1991	
Wasamac	Beauchastel Twp.	Wright-Harqueaves	1965-1971	2.3 tonnes @ 5.24 gpt Au, 0.75 gpt Ag
Remnor	Rouyn Twp.	Noranda	1985-1989	
Heva	Joannes Twp.	Heva Gold	1951-1952	36,000 tonnes @ 6.43 gpt Au
Don Chester	Duparquet Twp.		1933-1956	
Duqesne	Destor Twp.		1949-1952, 1990-1991	
Lanaudiere	Destor twp.	New Thurbois	1983-	

ONTARIO

MINE NAME	LOCATION	PRODUCER	MINE LIFE SPAN	PRODUCTION
Renabie	Missanabie	Renabie-Barrick	1947-1979, 1981-1991	5.6 mil. tons @ 0.20 opt Au
Goldlumb	Echo Twp.		1982-1985	111,891 tons @ 0.15 opt Au
MATHESON AREA				
Centre Hill	Munro Twp.		1967-1970	327,007 tons @ 0.01 opt Au
Canadian Arrow	Hislop Twp.		1980-1983	249,593 tons @ 0.06 opt Au
Ross	Hislop Twp.	Hollinger	1936-1989	6.7 mil. tons @ 0.15 opt Au
Hislop East	Hislop Twp.		1990-1994, 1999-2000	388,724 tons @ 0.125 opt Au
Holt-McDermott	Holloway Twp.	Camflo-Barrick	1988-2004	8.2 mil. tons @ 0.16 opt Au
Holloway	Holloway Twp.	Hemlo-Battle Mtn.-Newmont	1996-Present	4.7 mil. tons @ 0.16 opt Au
Glimmer	Hislop Twp.	Extall	1997-2001	1.2 mil. tons @ 0.17 opt Au
Clavos	Stock Twp.	St. Andrews	2006-2007	140,190 tonnes @ 3.28 gpt Au
Stock	Stock Twp.	St. Andrews	1989-1994, 2000-Present	815,400 tons @ 0.16 opt Au
Detour Lake	Lower Detour Lake	Placer	1983-1999	14.3 mil. tonnes @ 0.15 opt Au
Surluga	McMurray Twp.	Citidel	1968-1969, 1975-1981	87,460 tons @ 0.18 opt Au
O'Sullivan	O'Sullivan Lk.	Cons Louanna	1982-1984	64,000 tons @ 0.24 opt Au
Sapawe	McCaul Twp.	Atiko	1964-1966	33,013 tons @ 0.14 opt Au
Orofino	Scadding Twp.	Camflo	1987-1990	
LARDER LAKE CAMP				
Kerr	McGarry Twp.	Kerr Addison	1938-1996	40.3 mil. tons @ .26 opt Au
Cheminis	McVittie Twp.		1991-1996	179,013 tons @ 0.10 opt Au
Buffonta	Garrison Twp.		1981-1982, 1991-1992	117,013 tons @ 0.10 opt Au
Armistice	McGarry Twp.		1995-1997	8,282 tons @ 0.125 opt Au
Barber Larder	McGarry Twp.		1988-1989	30,000 tons @ 0.10 opt Au
Chesterville	McGarry Twp.		1930-1952	3.3 mil. tons @ 0.11 opt Au
Omega	McVittie Twp.	Kerr Addison Omega Gold	1926-28, 1936-1948	1.6 mil. tons @ 0.13 opt Au
KIRKLAND LAKE CAMP				
Toburn	Teck Twp.	Toburn	1917-1953	1.2 mil. tons @ 0.48 opt Au, .24 opt Ag
Sylvanite	Teck Twp.	Sylvanite GM	1927-1961	5.1 mil. tons @ 0.33 opt Au, .20 opt Ag
Kirkland Townsite	Teck Twp.		1958-1959	4,230 tons @ 0.45 opt Au, .09 opt Ag
Kirkland Lake	Teck Twp.		1919-1960	3.1 mil. tons @ 0.37 opt Au, .11 opt Ag
Lake Shore	Teck Twp.		1918-65, 82-87, 97-98	17.2 mil tons @ 0.50 opt Au, .23 opt Ag
Wright rgreaves	Teck Twp.		1921-1965	9.9 mil. tons @ 0.49 opt Au, .18 opt Ag
Teck-Hughes	Teck Twp.	Teck Corp.	1917-1968	9.6 mil.tons @ 0.38 opt Au, .14 opt Ag
Upper Canada	Gauthier Twp.	Inco	1938-1971	4.6 mil. tons @ 0.30 opt Au, .42 opt Ag
Upper Beaver	Gauthier Twp.	Upper Canada	1963-1985, 1967-1972	505,000 tons @1.0% Cu, 0.24 opt Au
McBean	Gauthier Twp.		1984-1986	557,000 tons @ 0.08 opt Au
Macassa	Teck Twp.	Lac Minerals-Kirkland Lake Gold	1933-1999	7.9 mil. tons @ 0.45 opt Au, .13 op Ag to 1994
Anoki	Gauthier Twp.		1987	28,000 tons @ 4.0 gpt Au
Bidgood	Lebel Twp.		1934-1951	580,367 tons @ 0.27 opt Au
Kirkland Lake Gold	Teck Twp.		2002-Present	

MINE NAME	LOCATION	PRODUCER	MINE LIFE SPAN	PRODUCTION
MATACHEWAN CAMP				
Matach Cons	Powell Twp.		1934-1956, 1980-1982	3.6 mil. tons @ .105 opt Au, .04 opt Ag
Midlothian	Midlothian Twp.	Stairs Gold Mines	1965-1966	15,835 tons @ 0.226 opt Au
Young-Davidson	Powell Twp.	Y.D.Gold Mines	1934-1957	6.1mil.tons @ .100 opt Au, .02 opt Ag
TIMMINS CAMP				
Hollinger	Tisdale Twp.	Hollinger	1906-1968, 1976-1988	65.8 mil. tons @ 0.29 opt Au
McIntyre	Tisdale Twp.	McIntyre	1912-1988	37.5 mil. tons @ 0.29 opt Au
Dome	Tisdale Twp.	Dome-Placer-Gold Corp/Kinross	1910-Present	100.1 mil. tons @ 0.156 opt Au
Paymaster	Tisdale Twp.	Paymaster	1915-1919, 1962-1966	5.6 mil. tons @ .250 opt Au
Coniaurum	Tisdale Twp.	Westfield	1913-1918, 1926-1961	4.5 mil. tons @ 0.25 opt Au
Buffalo Ankerite	Deloro Twp.	Buffalo Ankerite	1926-1953	5.0 mil. tons @ 0.19 opt Au
Aunor	Deloro Twp.	Noranda	1939-1984	8.5 mil. tons @ 0.30 opt Au
Delnite	Deloro Twp.	Sylvanite	1937-1964, 1987-1988	3.8 mil. tons @ 0.20 opt Au
Desantis	Ogden Twp.		1939-1942, 1961-1964	196,928 tons @ 0.18 opt Au
Naybob	Ogden Twp.	Kenilworth	1934-1966	304,100 tons @ .230 opt Au
Pamour	Whitney Twp.	Noranda	1936-1999, 2005-Present	48.6 mil. tons @ 0.07 opt Au
Broulan Reef	Whitney Twp.	Broulan Reef	1915-17, 1947-51, 1951-56	2.1 mil. tons @ 0.23 opt
Hallnor	Whitney Twp.	Noranda	1938-1968, 1961	4.2 mil. tons @ .381 opt Au, .03 opt Ag
Hoyle Pond	Whitney Twp.	Falconbridge, Porcupine J.V.	1941-1949, 1984-Present	6.1 mil. tons @ 0.38 opt Au
Owl Creek	Hoyle Twp.	Falconbridge	1981-1984	1.36 mil. tons @ 0.12 opt Au
Preston East Dome	Tisdale Twp.	Rio Algom	1939-1968	6.3 mil. tons @ .245 opt Au
Bon Whit	Whitney Twp.		1951-1954	200,555 tons @ 0.34 opt Au
Hugh Pam	Whitney Twp.		1926, 1948-1965	636,751 tons @ 0.19 opt Au
Davidson-Tisdale	Tisdale Twp.		1918-1920, 1988	53,221 tons @ 0.18 opt Au
Bonetal	Whitney Twp.		1941-1951	352,254 tons @ 0.15 opt Au
Broulan Porcupine	Whitney Twp.	Broulan Reef	1915-1965	1.15 mil. tons @ 0.31 opt Au
Bell Creek	Hoyle Twp.	Falconbridge	1987-1994	576,000 tons @ 0.19 opt Au
Marhill	Hoyle Twp.	Canamex	1989-1991	150,800 tons @ 0.199 opt Au
Night Hawk	Cody Twp.	Royal Oak	1996-1999	1.48 mil. tons @ 0.12 opt Au
Aquarius	MacKlem Twp.	Aquarius Porcupine	1984	137,000 tons @ .245 opt Au
PICKLE LAKE CAMP				
Central Patricia	Connell Twp.	Teck	1934-1951	1.7 mil. tons @ 0.36 opt Au
Pickle Crow	Connell Twp.	Highland-Crow	1935-1966	3.2 mil. tons @ 0.45 opt Au
Golden Patricia	Kapichi Lake Area		1988-1997	1.2 mil. tonnes @ 17.4 gpt Au
Dona Lake	Dona Lake Area	Placer	1989-1994	939,237 tonnes @ 0.26 opt Au
Musselwhite	Pickle Lake North	Placer-Gold Corp.	1997-Present	1.3 mil tonnes @ 5.7 gpt Au
GERALDTON CAMP				
McLeod-Cockshutt	Errington Twp.	McLeod-Cockshutt	1938-1970	11.2 mil. tons @ 0.14 opt Au
Little Long Lac	Errington Twp.	Little Long Lac	1934-1956	1.8 mil. tons @ 0.34 opt Au
Hard Rock	Errington Twp.	Lac	1938-1951	1.5 mil. tons @0.18 opt Au,.006 opt Ag
Leitch	Sturgeon River Area	Leitch Gold Mines	1936-1968	1.0 mil. tons @ 0.92 opt Au
Magnetic Cons	Errington Twp.		1938-1952	359,910 tons @ 0.42 opt Au,.019 opt Ag
Mosher	Geraldton	McLeod-Cockshutt	1962-1970	475,000 tons @ 0.11 opt Au,.007 opt Ag
Talmora	Geraldton	Tombil	1947-1948	6,034 tons @ 0.21 opt Au
Bankfield	Geraldton	Bankfield Cons.	1937-1947	231,000 tons @ 0.29 opt Au
HEMLO CAMP				
Golden Giant	Bomby Twp.	Hemlo-Newmont	1985-2005	19.2 mil. tonnes @ 10.64 gpt Au
David Bell	Bomby Twp.	Corona Teck	1985-2005	9.0 mil. tonnes @ 13.31 gpt Au
Mosher Longlac	Long Lac		1962-1966	2.7 mil. tons @ 0.12 opt Au
Northern Empire	Beardmore		1934-41, 1949	425,866 tons @ 0.34 opt Au
GOUDREAU CAMP				
Kremzar	ACR Lands	Canamac-Patricia	1988-1990	392,858 tons @ 0.10 opt Au
Magina	ACR Lands	Muscocho	1988-1992	168,679 tons @ 0.15 opt Au
Edwards	ACR Lands	Eagle River	1997-2001	389,309 tons @ 0.36 opt Au
Cline	ACR Lands	Cline Lake	1924-1942, 1947-1948	331,842 tons @ 0.14 opt Au

MINE NAME	LOCATION	PRODUCER	MINE LIFE SPAN	PRODUCTION
MISHIBISHU CAMP				
Magnacon	Mishibishu Lake Area	Muscocho-Eagle River	1989-1991	165,000 tons @ 0.04 opt Au
Eagle River	Mishibishu Lake Area	Eagle River, Wesdome	1995-Present	2.4 mil. tons @ 0.28 opt Au
Mishi Pit	Mishibishu Lake Area	Eagle River, Wesdome	2002-2004	96,000 tons @ 0.14 opt Au
RED LAKE CAMP				
Red Lake	Balmer Twp.	Gold Corp (Campbell-Dickenson)	2006-Onward	0.7 mil. tons @ .91 opt Au
Campbell	Balmer Twp.	Placer gold	1949-2006	19.9 mil. tons @ .56 opt Au
Dickenson	Balmer Twp.	Dickenson-Goldcorp	1948-1996, 2000-2005	9.6 mil. tons @ .62 opt Au
Cochenor-Williams	Dome Twp.	Cochenor-Willans	1939-1971	2.3 mil. tons @ .54 opt Au
McKenzie Island	Dome Twp.	McKenzie Red Lake	1935-1966	2.4 mil. tons @ .28 opt Au
MacMarmac	Dome Twp.	Marbey Mines	1940-1948	152,928 tons @ .30 opt Au
Wilmar	Dome Twp.	Wilmar	1967-1971	193,435 tons
Hasaga	Heyson Twp.	Little Long Lac	1938-1952	1.5 mil. tons @ .14 opt Au
Madsen	Heyson Twp.	Madsen Red/Bulord/Claude Res.	1936-1976, 1997-1999	8.7 mil tons.@ 0.28 opt Au
Starratt Olsen	Heyson Twp.	Starrett Nickel	1948-1956	907,813 tons @ .18 opt Au
Buffalo	Heyson Twp.	Wilanout Res.	1981-1982	31,986 tons @ .05 opt Au
Abino	Dome Twp.	Goldquest	1985-1986	2737 tons @ .51 opt Au
Lake Rowan	Todd Twp.	Goldquest	1986-1988	13,020 tons @ .10 opt Au
H.G.Young	Balmer Twp.	H.G. Young Mines	1960-1963	288,179 tons @ .19 opt Au
Berens River	Favourable Lake	Berens River Mines	1939-1948	560,607 tons @ .28 opt Au
Jason (Argosy)	Cassumit Lake	New Jason Mines	1934-1952	276,573 tons @ 0.37 opt Au

MANITOBA

MINE NAME	LOCATION	PRODUCER	MINE LIFE SPAN	PRODUCTION
			1948-1950	18,000 tons @ 0.82 opt Au
Nor-Acme	Snow Lake	Nor Acme-High Lake	1948-1958, 1995-2004	5.4 mil. tons @ 0.12 opt Au, .008 opt Ag
Donjon			1955-1957	
Squall Lake		Solidor	1984-1988	684,000 tons @ 3.4 gpt Au
Tarton Lake	Flin Flon	Granges	1987-1989	544,000 tons @ 7.23 gpt Au
Puffy Lake			1987-	
Burnt Timber				1.0 mil. tons @ 3.77 gpt Au
BISETT LAKE				
Jeep	Bissett Lk.		1947-1950	18,000 tons @ 26.4 gpt Au
Forty-Four	Bissett Lk.		1926-1968	
San Antonio	Bissett Lk.	San Antonio-Harmony-San Gold	1962-68, 82-83, 98-Present	6.6 mil. tons @ 8.8 gpt Au
Ogama Rockland	Bissett Lk.		1948-1958	155,862 tons @ 9.61 gpt Au, 1.1 gpt Ag
San Gold #1	Bissett Lake	San Gold	2006-Present	
LYNN LAKE				
BT (Key Stone)	Lynn Lake		1993-2000	
Farley Lake	Lynn Lake	Hudson bay	1996-	
MacLellan	Lynn Lake	Falconbridge		785,000 tons @ 5.96 gpt Au
Markwan Lake	Lynn Lake	Lynngold Resources		0.5 mil. tons @ 3.04 gpt Au

SASKATCHEWAN

MINE NAME	LOCATION	PRODUCER	MINE LIFE SPAN	PRODUCTION
New Cor	LaRonge	Flin Flon Mines	1947-1948	42,500 tons @ .313 opt Au, .88 opt Ag
Star Lake	LaRonge		1987-1989	50,500 tons @ .469 opt Au
Jolu	LaRonge	Mahogany Res.	1988-1991	50,500 tons @ .469 opt Au
Seebee	LaRonge	Claude Resources	1991-Present	
Contact Lake	Uranerz Lake	Cameco	1995-1996	587,000 tons @ 0.23 opt Au, 80 opt Ag
Komis	Waddy Lake	Waddy Lake Mines	1996-1997	129,200 tons @ 1.00 opt Au
Graham	Waddy Lake	Waddy Lake Mines	1996-1997	129,000 tons @ 0.22 opt Au
Jasper			1990-1991	148,000 tons @ 18.8 gpt Au

MINE NAME	LOCATION	PRODUCER	MINE LIFE SPAN	PRODUCTION

BRITISH COLUMBIA

MINE NAME	LOCATION	PRODUCER	MINE LIFE SPAN	PRODUCTION
Silbak Premier		Westmin	1918-1979	4.2 mil. tons @ 13.36 gpt Au, 302.12 gpt Ag, 0.12%Cu, 6.3%Pb, 2.10% Zn
Privateer	Zeballos		1934-1975	146,835 tonnes @ 36.10 gpt Au, 14.71 gpt Ag, .02% Cu
Polaris-Taku	NE Coast	Cominco	1938-1951	689,090 tons @ 10.45 gpt Au, .53 gpt Ag, .05% Cu
Bralorne-Pioneer	Bralorne	Bralorne-Mascot	1900-1949, 1971-1980	7.3 mil. tonnes @ 17.73 gpt Au, 4.06 gpt Ag
Scottie	Stewart	Northair Mines	1982-1985	160,264 tons @ 18.60 gpt Au, 10.14 gpt Ag
Baker	Toogodongood		1981-1983	81,878 tons @ 15.68 gpt Au, 290.84 gpt Ag
Blackdome	Clinton	Blackdome	1986-1990, 1998-1999	342,340 tons @ 21.86 gpt Au, 76.13 gpt Ag
Tillicum	Burton		1981-1993	5729 tons @ 28.80 gpt Au, 38.23 gpt Ag
Lawyers	Toogodongood	Cheni	1989-1992	619,869 tons @ 8.7 gpt Au, 18.26 gpt Ag
Premier Gold	Stewart	Westmin	1989-1996	2.2 mil. tons @ 2.76 gpt Au, 28.32 gpt Ag
Golden Bear	Telegraph Creek	Wheaton River	1990-1995, 1997-2002	1.3 mil. tons @ 8.13 gpt au, 1.69 gpt Ag
Johnny Mtn.	Terrace	Skyline Gold	1988-1990, 1993	227,247 tons @ 12.39 gpt Au, 19.14 gpt Ag, .49% Cu
Shasta	Smithers	Sable Res.	1989-2000, 2005	133,813 tons @ 4.62 gpt Au, 249.15 gpt Ag
Dome Mtn.	Smithers	Cdn. United	1991-1992	30,890 tons @ 11.72 gpt Au, 4.44 gpt Ag
Silver Butte	Stewart	Northair	1991-1993	108,391 tons @ 7.86 gpt Au, 23.45 gpt Ag
Eskay Creek	Iskut River	Corona-Homestake-Barrick	1995-Present	1.9 mil. tonne @ 53.41 gpt Au, 2545.58 gpt Ag
QR Mine	Quesnel	Kinross	1995-1996	1.0 mil. tons @ 3.69 gpt Au, 1.07 gpt Ag
Table Mtn.	Cassiar	Erickson, Cusac	1978-1999	634,800 tons @ 14.33 gpt Au, 4.39 gpt Ag
Evening Star	Rossland			1,434 tons @ 15.06 gpt Au
Snip	Iskut River	Cominco	1991-1999	1.4 mil. tons @ 23.86 gpt Au, 9.2 gpt Ag, .03% Cu
Caribou Gold	Quesnel	Mosquito Creek	1934-1987	2.7 mil. tonnes @ 14.19 gpt Au, 1.7 gpt Ag
Nickel Plate	Hedley	Homestake		15.0 mil. tonnes @ 5.1 gpt Au, 1.25 gpt Ag, .01% Cu
Northair	Pemberton		1974-1982	475,042 tons @ 10.91 gpt Au, 55.38 gpt Ag, .09% Cu
Ladner Creek	Hope	Carolin	1982-1984	901,566 tons @ 1.62 gpt Au, .12 gpt Ag
Elk	Peachland	Almaden	1992-1995	16,570 tons @ 91.67 gpt Au, 114.85 gpt Ag
Equity Silver	Houston	Placer Dome	1980-1994	32.6 mil. tons @ .48 gpt Au, 68.08 gpt Ag, .28% Cu
Sheep Creek	Salmo		1902-1988	2.3 mil. tons @ 9.24 gpt Au, 3.64 gpt Ag
Pinchi Lake	Ft. St.James	Cominco	1940-1975	

NORTHWEST TERRITORIES

MINE NAME	LOCATION	PRODUCER	MINE LIFE SPAN	PRODUCTION
YELLOWKNIFE CAMP				
Con	Yellowknife	Cominco-Miramar	1938-2003	11.5 mil. tonnes @ 15.6 gpt Au
Giant	Yellowknife	Falco-McWatters-Miramar	1948-2004	15.0 mil. tonnes @ 15.6 gpt Au
Discovery	Yellowknife		1949-	
Ptarmigan	Yellowknife	Cominco-Treminco	1941-1942, 1993-1996	
Tundra	Courageous Lk	Tundra Gold	1964-1968	104,474 oz Au
Salmita	Courageous Lk	Falconbridge	1983-1987	179,906 oz. Au

MINE NAME	LOCATION	PRODUCER	MINE LIFE SPAN	PRODUCTION
			NUNAVUT	
Lupin	Contwoyto Lake	Echo Bay-Kinross	1982-1998,2000-2003	8.9 mil. tonnes @ 9.52 gpt Au, 1.54 gpt Ag
Cullaton Lake	Indin Lake	Cullaton Lake-Neptune Res.	1981-85, 90-91, 94-97	373,000 tonnes @ 10.60 gpt Au to 1985
			YUKON	
Ketza River	Ross River		1988-1990	1.4 mil. tons @ 3.54 gpt Au
Brewery Creek	Dawson City		1996-2004	13.3 mil. tons @ 1.44 gpt Au
Mount Nansen	Carmacks		1996-1998	85,727 tons @ 14 gpt Au, 283 gpt Ag
			BASIC INTRUSIONS (COPPER/NICKEL/PGM)	
			NEWFOUNDLAND	
Voisey Bay	Nain,Lab.	Inco-CVRD	2006-Present	
			QUEBEC	
Raglan	Deception Bay	Falco-Xstrata	1997-Present	
Lorraine	Gaboury	McIntyre Mines	1964-1968	600,000 tonnes @ 1.08% Cu, 0.48% Ni
Lac Renzy	Hainault Twp.	Renzy Mines	1969-1972	766,000 tonnes @ 0.53% Cu, 0.48% Ni
Marbridge	LaMatte Twp.		1962-1968	
CHIBOUGAMAU CAMP				
Campbell Chi.	Obolski Twp.	Campbell-Chibougamau Mines	1955-Present	
Kokko Creek	McKenzie Twp.	Campbell Res.	1959-1975	751,000 tonnes @ 1.98% Cu, 0.31 gpt Au
Copper Rand	McKenzie Twp.	Northgate Patino	1962-	10.1 mil. tonnes @ 1.90% Cu, 2.20 gpt Au, 7.50 gpt Ag
Portage Is.	Roy Twp.	Northgate Patino	1960-	3.1 mil. tonnes @ 1.90% Cu, 2.51 gpt Au, 8.10 gpt Ag
Cooke Mine	Levy Twp.	Falconbridge	-1991	
Springer	Levy Twp.	Falconbridge	-1991	
Perry	Levy Twp.	Falconbridge	-1991	
Icon-Sullivan	O'Sullivan Twp.	Icon Syndicate	1967-1974	1.2 mil. tonnes @ 2.18% Cu
Bateman Bay	Lac Dore, Que.	Bateman Bay	1959-1960	15,000 tonnes @ 2.0% Cu, 0.46 opt Ag, 0.07 opt Au
Bruneau	McKenzie Twp.	Brueau	1966-1968	69,000 tonnes @ 1.52% Cu, 0.69 gpt Au, 15.09 gpt Ag
Chib-Kayband	Obalski Twp.	Campbell Res.	1965-1974	114,000 tonnes @ 1.3% Cu
Copper Cliff	McKenzie Twp.	Patino	1970-1974	864,000 tonnes @ 1.6% Cu, 0.60 gpt Au, 6.50 gpt Ag
Grandroy	Roy Twp.	Campbell Res	1967-1968, 1973-1975	390,000 tonnes @ 1.22% Cu, 0.68 gpt Au
Jaculet	McKenzie Twp.	Patino	1961-1971, 1974-1977	1.2 mil. tonnes @ 1.84% Cu, 1.44 gpt Au, 6.85 gpt Ag
Merrill Is.	Obalski Twp.	Campbell Res	1958-1975, 1977	2.6 mil. tonnes @ 1.46% Cu, 0.40 gpt Au
Chibougamau Goldfields	McKenzie Twp.	Patino	1963-1971	207,000 tonnes @ 1.78% Cu, 2.81 gpt Au
Cedar Bay	McKenzie Twp.	Camchib	1958-1987	4.2 mi. tonnes @ 1.58% Cu, 3.00 gpt Au
Henderson	Roy Twp.	Camchib	1959-1987	7.6 mil. tonnes @ 1.81% Cu, 1.48 gpt Au
Obalski	Obalski Twp.	United Obalski	1964-1965	75,000 tonnes @ 1.20% Cu, 2.60 gpt Au, 6.04 gpt Ag
Allard Lake	Sorel, Que	Quebec Iron	1949-	
Tio	Havre-St.Pierre, Que	QIT-Fer et Titane	-Present	
SUDBURY BASIN				
Creighton		Inco-CVRD	1901-Present	173.0 mil. tonnes @ 1.52% Ni, 1.22% Cu
Falconbridge	Garson Twp.	Falconbridge	1929-1983, -1991	36.4 mil. tons @ 0.89% Cu, 1.58% Ni
Levack West	Levack Twp.	Inco-FNX	1937-1997	60.0 mil. tons @ 1.31% Cu, 2.00% Ni, 1.5 gpt TPM

MINE NAME	LOCATION	PRODUCER	MINE LIFE SPAN	PRODUCTION
McKim	McKim Twp.	Falconbridge	1949-	
East Rim	Garson Twp.	Falconbridge	1951-1990	9.6 mil. tons @ 0.76% Cu, 1.15% Ni
Milnet		Falconbridge	1952-	
East		Falconbridge	1953-	
Hardy		Falconbridge	1954-	
Clarabelle #1		Inco	1961-1975	20.0 mil. tons
North Mine		Falconbridge	1965-	
MacLennan	MacLennan Twp.	Inco	1966-1972	2.1 mil. tons
Totten	Drury Twp.	Inco	1966-1970	1.0 mil. tons
Crean Hill		Inco	1964-72, 1973-74, 1986-?	13.1 mil. tons to 1986
Strathcona		Falconbridsge	1968-	
Longvack		Falconbridge	1968-	
Copper Cliff North		Inco-CVRD	1968-78, 1983 Present	17.6 mil. tons to 1986
Kirkwood	Garson Twp.	Inco	1914-1916,1969-1976	2.7 mil. tons @ 1.00% Cu, 0.90% Ni
Coleman	Levack Twp.	Inco-CVRD	1971-Present	
Little Stobie	Blezard Twp.	Inco-CVRD	1971-Present	13.4 mil. tons to 1986
Victoria	Denison Twp.	Inco	1914-1916, 1973-1978	1.54 mil. tons @ 2.26% Cu, 1.57% Ni, 2.0 gpt TPM
Boundary		Falconbridge	-1973	
Murray	McKim Twp.	Inco	1950-1971	35.2 mil. tons
McCready West	Levack Twp.	inco-FNX	1974-1999,2003-	15.8 mil. tons @ 1.70% Cu, 1.44% Ni, 1.5 gpt TPM to 1999
Lockerby		Falconbridge	1977-1998	
Upper Coleman	Levack Twp.	Inco	1970-1982	7.7 mil. tons
Clarabelle #2	McKim Twp.	Inco-CVRD	1967-Present	3.2 mil. tons to 1986
Frood	McKim Twp.	Inco-CVRD	1923-Present	192.8 mil. tons to 1986
Stobie	Blezard Twp.	Inco-CVRD	1942-Present	108.5 mil. tons to 1986
Victor	MacLennan Twp.	Inco	1960-	134,100 tons
Stracona		Inco	1978-	
Fraser		Falco-Xstrata	1981-Present	
Copper Cliff South		Inco-CVRD	1970-Present	15.6 mil. tons to 1986
Garson	Garson Twp.	Inco-CVRD	1907-1986,1994-Present	48.1 mil. tons to 1986
Whistle	Norman Twp.	Inco-FNX	1988-1991,1994-1997	5.7 mil. tons @ 0.3% Cu, 1.0% Ni, 0.034% Co
Onaping/Craig		Falco-Xstrata	1989-,1995-Present	
Lower Coleman		Inco	1991-	
Thayer Lindsley		Falco-Xstrata	1992-Present	
McCeady East		Inco-CVRD	1996-Present	
Mt. Nickel		Falconbridge		
Ellen Pit	Denison Twp.	Inco	1961-1971	1.2 mil. tons
Ajax	Temagami	Kanichee	1973-1976	278,263 tons @ 0.25% Ni, 0.56% Cu
Mount Nickel	Blezard Twp.		1953-	
Shebandowan	Hagey Twp.	Inco	1972-1985, 1995-1998	8.3 mil. tonnes @ 2.00% Ni, 1.20% Cu, 3.96 gpt PGM
Thierry	Pickle Lake	Umex	1976-1982	4.9 mil. tons @ 1.8% Cu, 0.17% Ni
Werner Lake	Werner lake	Cons Faraday	1962-1972	
Lac des Isles	Thunder Bay	North Amer. Paladium	1993-Present	32.6 mil. tonnes @ .06 opt Pd, .005 opt Pt, .004 opt Au, .04% Ni
TIMMINS CAMP				
Texmont	Bartlett Twp.	Texmont Mines	1971-1972	196,816 tons @ 0.96% Ni
Langmuir	Langmuir Twp.	Noranda	1974-1978,1991-1992	1.1 mil. tons @ 1.43% Ni
Redstone	Redstone River	McWatters-Liberty	1989-92, 1995-96, 2006-Present	324,000 tons @ 2.02% Ni
Montcalm	Montcalm Twp.	Falco-Xstrata	2004-Present	1.85 mil. tonnes @ 1.46% Ni, 0.7% Cu, 0.06% Co
Alexo/Kolex	Dundonald Twp.	Can. Arrow	2004-2005	17,400 tonnes @ 2.3% Ni, 0.23% Cu, 0.07% Co

MINE NAME	LOCATION	PRODUCER	MINE LIFE SPAN	PRODUCTION

MANITOBA

MINE NAME	LOCATION	PRODUCER	MINE LIFE SPAN	PRODUCTION
THOMPSON CAMP				
Marbridge			1962-	
Birch Tree	Thompson	Inco-CVRD	1969-1989, 1995-Present	
Moak	Thompson	Inco	1950-	4.5 mil. tons @ 0.71% Ni
Soab North	Thompson	Inco	1967-1971	900,000 tons @ 1.52% Ni
Pipe Pits	Thompson	Inco	1971-1984	18 mil. tons @
Manibridge	Thompson	Falconbridge	1971-1977	983,000 tons @ 1.81% Ni, 0.14% Cu
Thompson	Thompson	Inco-CVRD	1986-Present	25.0 mil. tons @ 2.97% NI
Thompson S	Thompson	Inco	1990-	
I-D Mine		Inco	2006-Present	
Bucko		Bowden Nickel		2.7 mil. tons @ 0.80% Ni
Discovery		Bowden Nickel		4.5 mil. tons @ 1.00% Ni
Maskwa West	Dunbarton	Falconbridge	1975-1976	1.2 mil. tons @ 1.28% Ni, 0.23% Cu
Dunbarton	Dunbarton	Cdn. Faraday	1964-1974	1.2 mil. tons @ 0.8% Ni, 0.3% Cu
LYNN LAKE CAMP				
Lynn Lake	Lynn Lake	Sherritt Gordon	1953-1976	37.6 mil. tonnes @ .22% Cu, 1.28% Ni
EL & A	Lynn Lake	Sherritt Gordon	1954-	1.0 mil. tons @ 1.0% Cu, 2.4% Ni
Namew Lake	Lynn Lake	Hudson Bay	1988-1993	2.4 mil. tons @ .70% Cu, 2.0% Ni

SASKATCHEWAN

MINE NAME	LOCATION	PRODUCER	MINE LIFE SPAN	PRODUCTION
Rottenstone	La Ronge		1965-1968	26,724 tons @ 3.2% Ni, 1.8% Cu, 0.47 opt TPM

BRITISH COLUMBIA

MINE NAME	LOCATION	PRODUCER	MINE LIFE SPAN	PRODUCTION
Giant Mascot	Hope	Giant Mascot Mines	1958-1974	4.2 mil. tons @ .35% Ni, .70% Cu

YUKON

MINE NAME	LOCATION	PRODUCER	MINE LIFE SPAN	PRODUCTION
Wellgreen	Kluane Lake	Hudson Bay	1972-1973	189,211 tons @ 2.23% Ni, 1.39% Cu

NUNAVUT

MINE NAME	LOCATION	PRODUCER	MINE LIFE SPAN	PRODUCTION
North Rankin	Rankin Inlet	Rankin Nickel	1957-1962	453,982 tonnes @ 3.3% Ni, 0.8% Cu, 1.02 gpt Pt, 2.05 gpt Pd

KIMBERLITES — DIAMONDS

NORTHWEST TERRITORIES

MINE NAME	LOCATION	PRODUCER	MINE LIFE SPAN	PRODUCTION
LAC DE GRAS CAMP				
Ekati	Lac de Gras	BHP, Diamet	1998-Present	33.0 mil. tonnes @ 1.0 carats/tonne
Diavik	Lac de Gras	RTZ,Aber Res.	2003-Present	7.1 mil. tonnes @ 3.7 carats/tonne

NUNAVUT

MINE NAME	LOCATION	PRODUCER	MINE LIFE SPAN	PRODUCTION
Jerico	Kitikmat Region	Tahera, Teck Corp.	2006-Present	400,000 tonnes @ .81 carats/tonne

MINE NAME	LOCATION	PRODUCER	MINE LIFE SPAN	PRODUCTION

IRON

NEWFOUNDLAND

MINE NAME	LOCATION	PRODUCER	MINE LIFE SPAN	PRODUCTION
Wabana	Bell Is.	Dosco	1892-1965	78.9 mil. tonnes @ 52% Fe
LABRADOR TROUGH (NFDL)				
Scully	Wabush	Cleveland Cliffs, Stelco, Dofasco	1965-Present	203 mil. tonnes conc. @ 66% Fe
Carol	Labrador City	Iron Ore Co.	1962-Present	71.0 mil. tonnes conc. @ 39% Fe
Knob Lake	Schefferville	Iron Ore Co.	1954-1982	120 mil. tonnes conc. @ 50-55% Fe

QUEBEC

MINE NAME	LOCATION	PRODUCER	MINE LIFE SPAN	PRODUCTION
LABRADOR TROUGH (QUEBEC)				
Ruth Lake	Wabush		1953-	
Gagnon 'C'	Wabush		1953-	
French	Knob Lake		1955-	
Gill	Knob Lake		1956-	
Farriman #1	Knob Lake		1957-	
Lac Jeannine	Wabush	Que.Cartier	1961-1977	
Wishart	Schefferville	Iron Ore Co.	1960-1982	
Wabush	Wabush		1960-	
Smallwood	Carol Lake		1962-	
Que./Cartier	Mount Wright	Que.Cartier	1975-	
Mt. Wright	Mount Wright	Que.Cartier	1986-Present	
Sidbec-Normines	Fire Lake		1976-	
Bristol	Shawville	Hilton Mines	1872-94,1956-76	
Chesbar	Desmaraisville		1970-1971	

ONTARIO

MINE NAME	LOCATION	PRODUCER	MINE LIFE SPAN	PRODUCTION
Marmoraton	Marmora	Bethlehem Steel	1955-1978	28.0 mil. tons @ 40% Fe
Moose Mtn.	Sudbury	National Steel	1907-11, 16-20, 59-78	
Adams	Boston Twp.	Jones & Laughlin	1964-1990	
Summit Lake	South Canonto Twp.	Tomelid	1995-	3.0 mil. tons @ 40% Fe
Griffith	Bruce Lake Area		-1968	
Sherman	Strathy Twp.	Cleveland Cliffs	1968-1989	
Nemegos	Chapleau		1973-	
WAWA CAMP				
Helen Mine	Wawa	Algoma	1900-1950	23.2 mil. tons @ 56% Fe
Sir James Dunn	Wawa	Algoma	1958-1967	8.1 mil. Tons @ 34% Fe
G.W. MacLeod	Wawa	Algoma	1960-1998	60.0 mil. tons @ 32% Fe
Lucy Pit	Wawa	Algoma	1968-1970	1.2 mil. tons @ 32.4% Fe
Victoria	Wawa	Algoma	1952-1960	10.2 mil. tons @ 33.7% Fe
GOUDREAU CAMP				
Rand Pit	ACR Lands	Algoma	1919, 1958-1959	70,000 tons @ 58% Iron Pyrite
Bear Claim	ACR Lands	Algoma	1958-1960	315,000 tons @ 30% Iron Pyrite
Goudreau A	ACR Lands	Algoma	1960-1961	225,000 tons @ 40% Iron Pyrite
Goudreau C	ACR Lands	Algoma	1914-1921, 1961	388,000 tons @ 40% Iron Pyrite
ATIKOKAN CAMP				
Hogarth, Errington, Roberts	Freeborn Twp.	Steep Rock Iron	1944-1979	79.0 mil. tonnes @ 56.5% Fe

MINE NAME	LOCATION	PRODUCER	MINE LIFE SPAN	PRODUCTION
Caland	Schwenger Twp.	Caland Ore Co.	1960-1979	43.6 mil. tonnes @ 55% Fe
Cdn. Charleson	Freeborn Twp.		1958-1965	6.7 mil. tonnes @ 12 % Fe

BRITISH COLUMBIA

MINE NAME	LOCATION	PRODUCER	MINE LIFE SPAN	PRODUCTION
Texada	Texada Island	Texada Mines	1952-1976	20.8 mil. tons @ 11.3% Fe, .05% Cu, .79 gpt Ag
Jedway	Moresby Island	Falconbridge	1962-1968	3.94 mil. tons @ 57.8% Fe
Tasu (Wesfrob)	Moresby Island	Falconbridge	1967-1983	16.2 mil. tons @ 17.2% Fe, .41% Cu, 3.11 gpt Ag

SILVER (SIMPLE LEAD-ZINC VEINS)

QUEBEC

MINE NAME	LOCATION	PRODUCER	MINE LIFE SPAN	PRODUCTION
Candego	Gaspe	Valmont Mining	1948-1954	70,000 tonnes @ 4.3% Zn, 6.4% Pb, 5.19 opt Ag
Anacon	Portneuf Co.		1948-	

NEW BRUNSWICK

MINE NAME	LOCATION	PRODUCER	MINE LIFE SPAN	PRODUCTION
Nigadoo	Beresford		1968-1977	

ONTARIO

MINE NAME	LOCATION	PRODUCER	MINE LIFE SPAN	PRODUCTION
Silver Mountain	Fort William		-1949	
Long Lake	Parham	Lynx Can.Explor.	1973-1974	

BRITISH COLUMBIA

MINE NAME	LOCATION	PRODUCER	MINE LIFE SPAN	PRODUCTION
Monarch	Field		1988-1957	822,010 tons @ 30.56 gpt Ag, 6.18% Pb, .954% Zn
Paradise	Inverness		1901-1952	66,760 tons @ 343.45 gpt Ag, 11.94% Pb, 5.97% Zn
Reeves MacDonald	Trail		1949-1971	5.8 mil. tons @ 3.42 gpt Ag, 4.75 gpt Au, 1.53% Pb, 3.86% Zn
Blue Bell	Kootenay Lk.	Cominco	1952-1982	4.8 mil., tons @ 46.04 gpt Ag, .065% Cu
Indian Mines	Portland Canal		1947-1953	12,803 tons @ 120.28 gpt Ag, 4.86% Pb, 6.08% Zn
Kootenay Belle	Kootenay Lk.		1904-1967	252,310 tons @ 13.9 gpt Au, 5.17 gpt Ag, .02% Pb, .02% Zn
HB Mine	Salmo		1955-1978	6.6 mil. tons @ 4.46 gpt Ag, .83% Pb, 4.34% Zn
Silvana	New Denver		1970-1988	499,303 tons @ 486 gpt Ag, 6.32% Pb, 5.79% Zn
Beaver Dell	Rock Creek	Teck Corp.	1915-1991	1.2 mil. tons @ 846.67 gpt Ag, .43 gpt Au
Torbrit		Dolly Varden	1928-1959	1.3 mil. tons @ 446.15 gpt Ag, 4.10% Pb

YUKON

MINE NAME	LOCATION	PRODUCER	MINE LIFE SPAN	PRODUCTION
United Keno Hill	Elsa	Falconbridge/Conwest	1946-1989	856,382 tons @ 1026.0 gpt Ag, 4.8% Pb, 3.9% Zn
Venus	Carmacks	Falconbridge	-1985	61,689 tons @ 305.14 gpt Ag, 10.98 gpt Au, 2.5% Pb, 1.5% Zn

NORTHWEST TERRITORIES

MINE NAME	LOCATION	PRODUCER	MINE LIFE SPAN	PRODUCTION
Terra	Camsell River	Terra Mines	1970-1985	520,500 tons @ 27.8 opt Ag
Norex	Camsell River	Terra Mines	1972-1983	69,190 tons @ 33.5 opt Ag
Silver Bay	Camsell River	Procon	1976-1977	3752 tons @ 60.0 opt Ag

MINE NAME	LOCATION	PRODUCER	MINE LIFE SPAN	PRODUCTION
			YUKON	
United Keno Hill	Elsa	Falconbridge/Conwest	1946 1989	856,382 tons @ 1026.0 gpt Ag, 4.8% Pb, 3.9% Zn
Venus	Carmacks	Falconbridge	-1985	61,689 tons @ 305.14 gpt Ag, 10.98 gpt Au, 2.5% Pb, 1.5% Zn
			NORTHWEST TERRITORIES	
Terra	Camsell River	Terra Mines	1970-1985	520,500 tons @ 27.8 opt Ag
Norex	Camsell River	Terra Mines	1972-1983	69,190 tons @ 33.5 opt Ag
Silver Bay	Camsell River	Procon	1976-1977	3752 tons @ 60.0 opt Ag

SILVER (COMPLEX CONTACT DEPOSITS)

ONTARIO

MINE NAME	LOCATION	PRODUCER	MINE LIFE SPAN	PRODUCTION
COBALT CAMP				
Silver Miller			1947-1957	
Silverfield			1966-	
Torbit			1966	
Silver Cliff			1949-	
Cobalt Lode			1908-1954	
United Cobalt			1951-1957	
Roy Silver			1954-	
Deer Horn		Cadesky Assoc.	1949-1953, 1960-1969	
Glen Lake		Cadesky Assoc.	1962-	
Silver Summit		Cadesky-Silverfields	1962-1983	
Nipissing 407			1924-1926, 1963-1971	
Bailey			-1971	
Penn Canadian			1905-1921,1968-1971	
Patricia Silver		Cadesky	-1971	
Nipissing 96			-1972	
Agnico Eagle		Agnico Eagle	-1972	
Beaver Temiskaming			1978-1983	
Keeley Frontier			1908-13, 21-31, 61-66	7.04 mil. ounces Ag
Monsilvo			1905-09, 46-56, 62-67	
Buffalo			1905-1959	
LaRose			1906-1924, 1952-1957	
O'Brien			1905-1937, 1952-1967	
Violet			1907-25, 51-52, 63-64	
Colonial			1907-1937, 1950-1964	
Juno			1918-1922, 1952-1955	
Kerr Lake			1905-28, 53-60, 67-70	
Christopher			1905, 1954-1956	
Victory			1921-26, 50-56, 63-65	
Ophir			1910, 1952-1957	
Mayfair			1910, 1945-46, 61-62	
GOWGANDA CAMP				
Miller Lake	Miller Twp.	O'Brien-Siscoe	1910-1972	1.85 mil. tons @ 22.0 opt
Castle-Trethewey	Miller Twp		1949-1965	
Mann			1912-1924, 1968-1970	
Rustev			1910-13, 36-38, 64-66	
Bonsall			1909-1910, 1966-1969	
Morrison			1925-27, 30-36, 52-54	
Shane-Darragh			1908-1913, 1953-1955	

MINE NAME	LOCATION	PRODUCER	MINE LIFE SPAN	PRODUCTION
		BRITISH COLUMBIA		
Utica		Keremos	1967-	
Equity		Houston	1980-1984	
O.B.		Greenwood	1987-1989	
		NORTHWEST TERRITORIES		
Echo Bay	Port Radium	Echo Bay Mines	1964-1981	383,409 tonnes @ 2068.0 gpt Ag, 1.40% Cu

LEAD, ZINC (MVT) STRATA FORM DEPOSITS

MINE NAME	LOCATION	PRODUCER	MINE LIFE SPAN	PRODUCTION
		NEWFOUNDLAND		
Daniels Harbour	Daniels Harbour	Teck Corp.	1975-1990	6.8 mil. tons @ 7.8% Zn
		ONTARIO		
Long Lake	Olden Twp.		1897-1925, 1973-1974	
		NORTHWEST TERRITORIES		
Pine Point	Pine Point	Cominco	1964-1988	68.8 mil. tonnes @ 6.7% Zn, 2.9% Pb
		NUNAVUT		
Nanisivik	Baffin Island	Conwest-Breakwater	1986-2002	12.4 mil. tons @ 7.9% Zn, 0.4% Pb, 32.0 gpt Ag
Polaris	Little Cornwallis Is.	Cominco-Teck Cominco	1982-2002	17.0 mil. tonnes @ 13.1% Zn, 3.5% Pb
		YUKON		
White Horse Cop.	Whitehorse	Hudson Bay	1967-1981	10.3 mil. tons @ 1.50% Cu

URANIUM

MINE NAME	LOCATION	PRODUCER	MINE LIFE SPAN	PRODUCTION
		ONTARIO		
BLIND RIVER CAMP				
Pronto	Elliot Lake Area	Rio Algom	1956-1961, 1968-1990	2.1 mil. tons @ .100% U3O8
Quirke Lk.#1	Elliot Lake Area	Rio Algom	1956-1958	
Buckles	Elliot Lake Area	Rio Algom	1957-1960	
Can Met	Elliot Lake Area	Denison Mines	1957-1992	
Cons Denison	Elliot Lake Area	Denison Mines	1957-1970	
Nordic Lake	Elliot Lake Area	Rio Algom	1957-1964	2.8 mil. tons @ .116% U3O8
Stanrock	Elliot Lake Area	Denison Mines	1957-1960	
Lacnor	Elliot Lake Area	Rio Algom	1958-1964	2.4 mil. tons @ .106% U3O8
Milliken	Elliot Lake Area	Rio Algom	1958-1961, 1978-1990	7.2 mil. tons @ .100% U3O8
Panel	Elliot Lake Area	Rio Algom	1958-1959	
Spanish-American	Elliot Lake Area	Rio Algom	1956-1960, 1978-1996	205,000 tons @ .100% U3O8
Stanleigh	Elliot Lake Area	Rio Algom	1960-1970	
Peter	Elliot Lake Area	Rio Algom	1968-1990	
Quirke Lk #2	Elliot Lake Area	Rio Algom		
Agnew Lake	Espanola	Kerr Addison	1977-1983	3.1 Mil. tons @ .034% U3O8

MINE NAME	LOCATION	PRODUCER	MINE LIFE SPAN	PRODUCTION
BANCROFT CAMP				
Bicroft Centre Lake			1956-1963	2.6 mil. tons @ .09% U3O8
Faraday Madawaska			1957-1964, 1976-1982	5.0 mil. tons @ 0.10% U3O8
Grey Hawk			1957-1959	80,000 tons @ .07% U3O8
Canadian Dyno		Conwest	1957-1960	660,000 tons @ .06% U3O8

SASKATCHEWAN

MINE NAME	LOCATION	PRODUCER	MINE LIFE SPAN	PRODUCTION
BEAVER LODGE CAMP				
Ace-Fay-Verna	Uranium City	Eldorado	1953-1981	3.9 mil. tons @ 0.20% U3O8
Gunnar	Uranium City	Gunnar Mines	1955-1963	5.0 mil. tonnes @ .17% U3O8
Eldorado-Hab	Uranium City	Eldorado		763 tonnes U3O8
Rix Smitty	Uranium City	Eldorado		436 tonnes U3O8
Cinch Lake	Uranium City	Lake Cinch Mines		285 tonnes U3O8
Cayzor Athabasca	Uranium City	Cayzor Mines	1957-1958	208,000 tonnes @ .33% U3O8
Eldorado-Dubyna	Uranium City	Eldorado		163 tonnes U3O8
Lorado	Uranium City	Eldorado		89 tonnes U3O8
Eldorado-Eagle	Uranium City	Eldorado		85 tonnes U3O8
Rix Leonard	Uranium City	Eldorado		77 tonnes U3O8
Nicholson	Uranium City	Eldorado	1950-55, 1959-2000	41 tonnes U3O8
National Exploration	Uranium City	Eldorado	1955	30 tonnes U3O8
Nesbitt Labine	Uranium City	Eldorado		23 tonnes U3O8
Eldorado-Fish Hook	Uranium City	Eldorado		15 tonnes U3O8
Eldorado-Martin Lake	Uranium City	Eldorado		11 tonnes U3O8
Uranium Ridge	Uranium City	Eldorado		10 tonnes U3O8
ATHABASCA BASIN CAMP				
Rabbit Lake	Rabbit Lake	Cameco Corp.	1975-1984	6.0 mil. tonnes @ .34% U3O8
Collins Bay A	Collins Bay	Cameco Corp.	1996-1997	
Collins Bay B	Collins Bay	Cameco Corp.	1985-1991	3.0 mil. tonnes @ .71% U3O8
Raven		UEX Corp.		
Horseshoe		UEX Corp.	1994-99, 2002-05	
Eagle Point		Cameco Corp.		
Midwest	McMahon Lake	Areva Redstone Res. Inc. Tenwest Uranium Ltd. Ourd (Canada) Co. Ltd.		
McClean Lake:JEB		Areva Denison Mines Ltd. Ourd (Canada) Co. Ltd.	1995-1997	
McClean Lake:Sue A,B,C,D Zones		Areva Denison Mines Ltd. Ourd (Canada) Co. Ltd.	1998-	
Key Lake	Key Lake	Cameco Corp. Areva	1983-1997	3.4 mil. tonnes @ 2.87 U3O8
Cluff Lake	Carswell Dome	Cogema Res. Inc.	1979-2002	
McArthur River	Key Lake	Areva Cameco	1999-	

NORTHWEST TERRITORIES

MINE NAME	LOCATION	PRODUCER	MINE LIFE SPAN	PRODUCTION
Ray Rock	Dogrib	Ray Rock Mines	1955-1959	80,041 tonnes @ 0.30% U3O8

MINE NAME	LOCATION	PRODUCER	MINE LIFE SPAN	PRODUCTION
PORPHYRIES (COPPER, MOLY, TIN)				
NEWFOUNDLAND				
BAIE VERTE AREA				
Ming West		Cons Rambler	1995-1996	110,000 tons @ 0.37% Zn, 5.6% Cu, .07 opt Au, .54 opt Ag
Ming		Cons Rambler	1971-1981	2.1 mil. tons @ 3.5% Cu, 2.2 g/t Au, 1.9 g/t Ag
East Mine		Cons Rambler	1967-1974	2.1 mil. tons @ 1.04% Cu
Little Deer Mine		Green Bay Mining	1974-1975	75,000 tons @ 2.3% Cu
Rambler		Cons Rambler	1961-1969	450,000 tons @ 2.16% Zn, 1.3% Cu, 4.66 gpt Au, 26.44 gpt Ag
Whalesback			1965-1972	3.8 mil. tons @ 1.0% Cu
Little Bay			1961-1969	2.6 mil. tons @ 0.8-2.0% Cu
Tilt Cove		M.J. Boylen	1957-1967	6.7 mil. tons @ 2.0% Cu, .006 opt Au
Gullbridge		M.J. Boylen	1967-1972	3.5 mil. tons @1.1% Cu
NOVA SCOTIA				
East Kemptville	Yarmouth Cty	Shell	1990-1992	10.0 mil. tons @ 0.21% Sn
NEW BRUNSWICK				
Mt. Pleasant	Fredericton	Billiton	1984-1985	Tin, Bismuth
George Lake	Fredericton	Cons Durham	1971-81, 1985-90	1.0 mil. tons @ 3.50% Sb
QUEBEC				
GASPE CAMP				
Copper Mtn.	Holland Twp.	Gaspe Copper	1955-1982	50.6 mil. tons @ 1.27% Cu
Needle Mtn.	Holland Twp.	Gaspe Copper	1968-1981	93.8 mil. tons @ 0.48% Cu
E-Zone	Holland Twp.	Gaspe Copper	1986-1999	6.2 mil. tons @ 2.60% Cu
Madeleine	Boisbuisson Twp.	Frobex-Teck	1969-1976, 1979-1982	7.4 mil. tonnes @ 1.09% Cu, 0.20 opt Ag
MALARTIC CAMP				
Cadillac Moly	Preissac Twp.	Lac Minerals	1943, 1965-70	1.9 mil. tonnes @ 0.14% MoS2, 0.03% Bi
Preissac	Preissac Twp.	Dumagami	1964-1971	2.3 mil. tonnes @ 0.19 MoS2, 0.03% Bi
Lacorne Moly	LaCorne Twp.	LaCorne Moly Co.	1942-1972	3.8 mil. tonnes @ 0.20% MoS2, 0.03% Bi
Quebec Lithium	LaCorne Twp.		1956-1965	
East Sullivan	Bourlamaque Twp.	East Sullivan	1948-1966	14.9 mil. tonnes @ 0.71% Zn, 1.04% Cu
Dunraine	Louvicourt Twp.	Dunraine Mining	1956-1958	258,000 tonnes @ 1.50% Cu, 0.21 gpt Au, 4.18 gpt Ag
Joliet	Rouyn Twp.	Joliet Que.	1949, 1955-1974	2.1 mil. tonnes @ 1.00% Cu
Main Oka	Oka, Que.	St. Lawrence Colum.	1961-1976	6.8 mil. tons @ 0.44% Cb205
Niobec	Chicoutimi, Que.	Cambior-IAM Gold	1976-	
ONTARIO				
Ryan Lake	Powell Twp.	Geo-Pax	1948-57, 1963-64	184,790 tons @ 2.0% Cu, 0.74% MOS2
Mattarow	Matachewan			
Temagami Copper	Temagami Lake	Copperfields-Teck	1956-1972	
Tribag	Batchewana Bay	Teck	1967-1973	1.10 mil. tons @ 1.65% Cu
Copper Corp	Batchewana Bay	Sheridan Geophysics	1965-1972	1.02 mil. tons @ 1.16% Cu
North Coldstream	Burchell Lake	Noranda	1906-1967	2.7 mil. tons @1.90% Cu, .16 opt Ag, .008 opt Au
MANITOBA				
Bernic Lake	Bernic Lake	Tanco	1969-1982	3.8 mil. tons @ 2.5% Li02, 3.4 mil. tons @ .13% Ta205

MINE NAME	LOCATION	PRODUCER	MINE LIFE SPAN	PRODUCTION
SASKATCHEWAN				
Anglo-Rouyn	Lac La Ronge	Rio Algom	1960-1972	1.7 mil. tons @ 1.71% Cu, 1.29 g/t Au, 5.5 g/t Ag
BRITISH COLUMBIA				
KAMLOOPS CAMP				
Afton-Ajax	Kamloops	Teck Corp.	1994-1997	7.8 mil. tons @ 3.83% Cu, .28 gpt Au, .54 gpt Ag
Afton	Kamloops	Teck Corp.	1978-1991	33.5 mil. tons @ .76% Cu, .44 gpt Au, 2.56 gpt Ag
Alwim (OK)	Kamloops	Dekalb	1980-1982	233,076 tons @ 1.77% Cu
Highmont	Kamloops	Teck Corp.	1981-1984	32.7 mil. tons @ .17% Cu, .02% MOS2
Big Bonanza			1971-1974	520,000 tons @ 1.5% Cu, .39 opt Ag, .008 opt Au
Rocher Deboule	Smithers	Arbor Res	1915-1954	36,450 tons @ 7.71% Cu, 3.68 gpt Au, 59.48 gpt Ag
Magnum	Fort St. John	Teck	1970-1975	498,132 tons @ 3.24% Cu
HIGHLAND VALLEY CAMP				
Highland Bell	Logan Lake	Teck	1949-	]
Bethlehem	Logan Lake		1962-1981	] 1,250 mil. tons @ .41% Cu, .007% MOS2
Valley Copper	Logan Lake	Teck-BHP-Highmount	1981-1986,1999-Present	]
Lornex	Logan Lake	RTZ		]
Brenda	Peachland	Brenda Mines	1968-1990	181.7 mil. tons @ .17% Cu, .04% MOS2
Old Sport	Port Alice	Benson Lake	1962-1973	2.7 mil. tons @ 1.73% Cu
Island Copper	Port Hardy	BHP Billiton	1971-1995	366.7 mil. tons @ .37% Cu, .01% MOS2, .10 gpt Au, .80 gpt Ag
Granduc	Stewart	Newmont-Esso	1971-1998,1981-1984	15.6 mil. tons @ 1.34% cu, 7.95 gpt Ag
Sunro	Vancouver Is.	Sunro Mines	1962-1978	1.3 mil. tons @ 1.16% Cu
PRINCETON CAMP				
Phoenix	Princeton	Granby	1900-1978	13.0 mil. tons @ 1.99% Cu, 2.18 gpt Au, 14.08 gpt Ag
Ingerbelle	Princeton	Granby-Newmont	1917-1996	185.5 mil. tons @ .51% Cu, 12.31 gpt Au, 158.62 gpt Ag
GRANISLE CAMP				
Bell Copper	Granisle	Noranda	1972-1982, 1985-1992	77.2 mil. tons @ .43% Cu, .17 gpt Au, .50 gpt Ag
Granisle	Granisle	Noranda	1966-1982	52.7 mil. tons @ .45% Cu, .13 gpt Au, 1.32 gpt Ag
Texada	Texada Is.		1899-1983	790,177 tons @
Tasu	Moresby Is.	Falconbridge	1966-1987	22.7 mil. tons @ .28% Cu, 2.32 gpt Ag
Huckleberry	Kemano	Imp.-Dowa-Furukaw	1997-Present	36.0 mil. tons @ .95% Cu, .01% MOS2
WILLIAMS LAKE CAMP				
Gibralter		Placer-Taseko/Ledcor	-1980, -1998, 2004-Present	315.2 mil. tons @ .32% Cu, .31% MOS2
Mt. Polley		Imp. Metals	1997-Present	50.0 mil. tons @ .22% Cu
Britannia	Howe Sound	Anaconda	1905-1977	47.4 mil. tons @ 1.20% Cu
Craigmont	Merritt	Craigmont	1961-1988	34.8 mil. tons @ 1.27% Cu
Kemess	North Central	Northgate	1998-Present	145.0 mil. tons @ .20% Cu
Kitsault	Alice Arm	Amax/Kennecott	1967-1982	13.4 mil. tons @ 0.11 MoS2
Boss Mtn.	Lac Lachache	Noranda	1965-1983	7.8 mil. tons @ 0.22% MoS2
Endako	Burns Lake	Placer-Black Pearl/Sojitz Moly	1965-Present	292 mil. tons @ 0.08% MoS2
Coxey		Rossland	1966-1970	939,397 tons @ 0.20% MoS2
Mt. Copeland	Revelstoke	Noranda	1970-1973	169,729 tons @ 0.77% MoS2

MINE NAME	LOCATION	PRODUCER	MINE LIFE SPAN	PRODUCTION
		NORTHWEST TERRITORIES		
Pit Zone	Flat River	Canada Tungsten	1962-1973	1.69 mil. tonnes @ 1.75% Wo3
E-Zone	Flat River	Canada Tungsten	1974-86, 2002-Present	
		YUKON		
Cantung	Watson Lake	Canada Tungsten	1962-86, 2002-Present	262,212 tons @ 1.05% Wo3 (2005)
		NOVA SCOTIA		
Mindamar	Stirling	Keltic Mining	1952-1956	1.6 mil. tons @ 6.3% Zn, 1.5% Pb, 0.80% Cu, 0.03 opt Au, 2.0 opt Ag
Gays River	Gays River	Esso-Hudbay-Acadian Gold		800,000 tons @ 4.0% Zn, 4% Pb
Yava Mines	Cape Breton	Yava Mines	1979-1981	388,000 tons @ 4.7% Zn
		ONTARIO		
New Calumet	Calumet Is.	New Calumet Mines	1943-1968	3.7 mil. tons @ 6.0% Zn, 1.7% Pb
Jardun	Sault Ste. Marie		1951-1957	130,536 tons @ 3.1% Zn, 4.4% Pb, 1.42 opt Ag
		BRITISH COLUMBIA		
Mineral King	Invermere		1950-1974	2.2 mil. tons @ 4.52% Zn, 1.87% Pb, .03% Cu, 26.24 gpt Ag
Sullivan	Kimberley	Cominco	1895-2001	150.5 mil. tons @ 5.81% Zn, 6.12% Pb, 61.56 gpt Ag
		YUKON		
Faro	Faro	Cyprus Anvil-Curragh Res.	1969-1982, 1986-1992	
Vangorda	Faro.	Cyprus Anvil-Curragh Res.	1991-1992, 1995-1998	165,000 tons @ 3.27% Pb, 4.14% Zn, 43.0 gpt Ag
Grum	Faro	Anvil Range-Curragh Res.	1995-1998	1.6 mil. tons @ 3.56% Pb, 5.34% Zn, 58.0 gpt Ag, .83 gpt Au
Sa Dena Has	Watson Lake	Teck	1991-1992	605,700 tonnes @ 5.25% Pb, 6.78% Zn

BASE METALS (VMS) STRATA BOUND DEPOSITS

MINE NAME	LOCATION	PRODUCER	MINE LIFE SPAN	PRODUCTION
		NEWFOUNDLAND		
Buchans	Millertown	Asarco	1928-1984	16.2 mil. tons@14.51% Zn, 7.56% Pb, 1.33% Cu, 1.37 g/t Au, 126.0 g/t Ag
		NEW BRUNSWICK		
BATHURST CAMP				
Keymet	Gloucester Cty.	Key Anacon	1954-1955	54,681 tonnes @ 4.3%Zn,4.7% Pb,0.66%Cu,48.0 gpt Ag
Brunswick #12	Gloucester Cty.	M.J. Boylen-Noranda-Xstrata	1964-Present	118.9 mil. tonnes @ 8.84% Zn, 3.48% Pb, 0.36% Cu, 102.6 gpt Ag
Brunswick #6	Gloucester Cty.	M.J. Boylen-Noranda	1966-1983	12.2 mil. tonnes @ 5.43% Zn, 2.15% Pb, 0.40% Cu, 67.0 gpt Ag
Heath Steele	Restigouche Cty.	Amax-Noranda	1957-1983, -1999	23.2 mil. tonnes@5.05% Zn, 1.75% Pb, 0.95% Cu, 66.6 gpt Ag to 1998
Heath Steele (oxides)	Restigouche Cty.	Noranda		170,000 tonnes @ 175.5 gpt Ag, 4.77 gpt Au
Wedge	Restigouche Cty.	Cominco	1962-1968	1.5 mil. tonnes @1.61% Zn, 0.65% Pb, 2.88% Cu, 20.6 gpt Ag
Caribou	Gloucester Cty.	Anaconda-Breakwater	1970, 74, 88-90, 97-98	1.34 mil. tonnes @ 6.78% Zn, 3.24% Pb, 0.32% Cu, 97.0 gpt Ag to 1998
Caribou (oxides)	Gloucester Cty.	Breakwater		61,500 tonnes @ 171.4 gpt Ag, 5.35 gpt Au
Murray Brook	Restigouche Cty.	Nova Gold	1980-, 1992-	1.01 mil. tonnes @ 61.4 gpt Ag, 1.79 gpt Au
Stratmat/NS Heath Steele	Restigouche Cty.	Cominco	1989-1993	1.14 mil. tonnes @ 8.11% Zn, 2.98% Pb, 0.35% Cu, 44.0 gpt Ag
Captain North Ext.	Restigouche Cty.	Noranda	1991-	39,000 tonnes @9.97% Zn, 4.42% Pb, 134.7 gpt Ag
Restigouche	Restigouche Cty.	New Jersey Zinc		230,000 tonnes @ 6.34% Zn, 5.49% Pb, 132.9 gpt Ag

MINE NAME	LOCATION	PRODUCER	MINE LIFE SPAN	PRODUCTION

QUEBEC

MINE NAME	LOCATION	PRODUCER	MINE LIFE SPAN	PRODUCTION
NORANDA CAMP				
Quemont	Rouyn Twp.	Noranda Mines	1949-1971	13.9 mil. tonnes @ 1.3% Cu, 2.4%Zn, 5.38 gpt Au, 30.9 gpt Ag
Horne	Rouyn Twp.	Noranda Mines	1927-1978, 1994, 1999	54.0 mil. tonnes @ 2.2% Cu, 5.83 gpt Au, 19.2 gpt Ag
Don Rouyn	Rouyn Twp.		-1949	
Mines Gallen	Dufresnoy Twp.	Noranda-Les Mines Gallen	1955-59, 81-85, 97-98	1.0 mil. tonnes@3.45% Zn, 196,000 tonnes@4.26% zn, 1.01 gpt Au, 31.3 gpt Ag
Mobrun	Dufresnoy Twp.		1987-1992	
Ansil	Duprat Twp.	Falconbridge Copper	1989-1992	
Bouchard-Hebert	Clericy Twp.	Cambior-Breakwater	1986-1992, 1995-2005	
Lac Dufault	Dufresnoy Twp.	Falconbridge Copper		
Corbet	Dufresnoy Twp.	Falconbridge Copper	1979-1986	
Millenbach	Dufresnoy Twp.	Falconbridge Copper	1970-1981	3.4 mil. tonnes @4.37% Zn, 3.45% Cu, 0.90 gpt Au, 47.81 gpt Ag
Vauze	Dufresnoy Twp.	Falconbridge Copper	1961-1965	350,000 tonnes @ 1.44% Zn, 3.19% Cu, 8.12 gpt Au
East Waite	Dufresnoy Twp.	Falconbridge Copper	1951-1961	1.4 mil. tonnes @ 4.62% Zn, 4.62% Cu, 1.37 gpt Ag
Norbec	Dufresnoy Twp.		1964-1978	3.7 mil. tonnes @ 4.86% Zn, 2.80% Cu
Amulet F	Dufresnoy Twp.		1930-1937, 1944-1962	254,000 tonnes @ 4.62% Zn, 4.62% Cu. 1.37 gpt Ag
Amulet A	Dufresnoy Twp.		1937-1962	4.8 mil. tonnes @ 4.62% Zn, 4.62% Cu, 1.37 gpt Au
Deldonna	Rouyn Twp.		1951-1952	
Delbridge	Rouyn Twp.	Delbridge Mining	1969-1971	360,000 tonnes @10.11% Zn, 0.45% Cu, 2.75 gpt Au, 109.34 gpt Ag
New Insco	Hebecourt Twp.		1975-1977	
Old Waite	Dufresney Twp.		1928-1949	1.1 mil. tonnes @4.62% Zn, 4.62% Cu, 1.37 gpt Ag
Joliet	Rouyn Twp.		1949, 1955-1974	
MATAGAMI CAMP				
Mattagami Lake	Galinee Twp.	Matagami Lake-Noranda	1963-1988	25.6 mil. tonnes @ 8.2% Zn,0.56% Cu,20.9 g/t Ag,0.41 g/t Au
Norita	Galinee Twp.	Matagami Lake-Noranda	1976-1996	3.9 mil. tonnes @ 3.9% Zn, 1.83% Cu, 41.4 g/t Ag, 0.51 g/t Au
Isle Dieu	Isle Dieu Twp.	Matagami Lake-Noranda	1989-1997	3.1 mil. tonnes @ 17.9% Zn, 1.01% Cu, 76.6 g/t Ag, 0.46 g/t Au
Orchan	Galinee Twp.	Orchan Mines-Noranda	1962-1982	4.5 mil tonnes @ 9.8% Zn, 1.02% Cu, 37.0 g/t Ag, 0.51 g/t Au.
Bell Allard South	Isle Dieu Twp.	Matagami Lake-Noranda	1968-1970	.23 mil. tonnes @ 9.2% Zn, 1.14% Cu, 37.0 g/t Ag, 0.51 g/t Au
Bell Allard	Isle Dieu Twp.	Noranda	1990-2004	3.6 mil. tonnes @ 13.7% Zn, 1.25% Cu, 40.6 g/t Ag, 0.9 g/t Au
Norita East	Galinee Twp.	Noranda	1976-1997	1.1 mil. tonnes @ 10.2% Zn, 0.80% Cu, 41.42 g/t Ag, 0.74 g/t Au
New Hosco	Daniel Twp.	New Hosco Mines	1963-1970	1.8 mil. tonnes @ 1.7% Zn, 1.73% Cu, 10.29 g/t Ag, 0.34 g/t Au
Garon Lake	Galinee Twp.	Matagami Noranda	1968-1970	0.47 mil. tonnes @ 2.2% Zn, 1.46% Cu, 10.29 g/t Ag, 0.34 g/t Au
Radiore #2	Galinee Twp.	Radiore Mines	1979-1980	0.14 mil. tonnes @ 1.3% Zn, 1.57% Cu, 8.57 g/t Ag, 0.31 g/t Au
JOUTEL CAMP				
Joutel Copper	Joutel Two.	Joutel Copper	1967-1975	0.4 mil. tonnes @ 8.9% Zn, and 1.3 mil. Tonnes @2.16% Cu
Poirier	Poirier Twp.	Rio Algom	1965-1975	2.0 mil. tonnes @ 1.84% Zn, 1.47% Cu, 7.34 gpt Au, 2.8 mil. Tonnes @ 2.34% Cu, 2.62 gpt Au
ABITIBI BELT				
Selbale	Brouillan Twp.	Selection Trust	1981-2004	44.0 mil. tonnes @ 7.0% Zn, 1.1% Cu, 0.60 gpt Au, 39.0 gpt Ag
Lemoine	Lemoine Twp.	Northgate-Patino	1975-1983	757,000 tonnes @ 9.52% Zn, 4.18% Cu
Barvue	Barraute Twp.		1952-1957	5.1 mil. tonnes @ 2.96% Zn, 38.05 gpt Ag
Abcourt	Barraute Twp.		1986-1990	
La Ronde #2		Agnico-Eagle Mines	1994-Present	0.96 mil. tonnes @2.5% Zn, 0.73% Cu, 7.3 gpt Au, 40.7 gpt Ag to 2002
La Ronde (Penna)		Agnico-Eagle Mines	2000-Present	4.34 mil. tonnes @ 4.4% Zn, 0.27% Cu, 5.0 gpt Au, 76.0 gpt Ag to 2002
Normet Mar	La Sarre		-1991	
Normetal	Desmeloizes Twp.	Mining Corp.	1925-1975	10.1 mil. tonnes @ 5.30% Zn, 0.79% Cu, 0.86 gpt Au, 65.48 gpt Ag
Louvicourt	Louvicourt Twp.	Aur-Noranda-Teck Cominco	1992-2005	
G.Langois	Grevet Twp.	Selco-Breakwater	1997-2000	
Coniagas	Lasueur Twp.	Coniagas-Falconbridge	1961-1967	652,000 tonnes @ 10.7% Zn, 1.0% Pb, 182.03 gpt Ag
Lynhurst	Destor Twp.	Lynhurst Mines	1956-1957	143,000 tonnes @ 1.83% Cu, 12.72 gpt Au
Hunter	Destor Twp.	Beattie-Duquesne	1957-1958	117,000 tonnes @ 1.0% Cu
Duvan		Aunore	1960	1,160 tonnes @ 12.0% Cu, 2.9 opt Ag
Candego	Boisbuisson Twp.	Candego Mines	1948, 1954	62,000 tonnes @ 3.45% Zn, 5.93% Pb, 172.2 gpt Ag
New Calumet	Grand Calumet Is.	Calumet Mines	1948-1968	3.4 mil. tonnes @ 6.06% Zn, 1.70% Pb, 0.65 gpt Au, 124.43 gpt Ag

MINE NAME	LOCATION	PRODUCER	MINE LIFE SPAN	PRODUCTION
Tetrault	Montauban Twp.	Key Anacon Mines	1913-1955	2.5 mil. tonnes @ 6.78% Zn, 2.27% Pb, 1.13 gpt Au, 131.30 gpt Ag
Montauban	Montauban Twp.	United Montauban	1953-	2.5 mil. tonnes @ 4.5% Zn, 1.5% Pb, 0.70 gpt Au, 85.0 gpt Ag
VAL D'OR CAMP				
East Sullivan	Bourlamaque Twp.	East Sullivan Mines	1949-1966	14.4 mil. tons @ 0.8% Zn, 1.0% Cu, 10.1 gpt Ag
Dunraine #1	Louvicourt Twp.	Agnico-Eagle	1956-1958	280,768 tonnes @ 1.78% Cu, 0.33 opt Ag, .001 opt Au
Louvem	Louvicourt Twp.	Soquem-Aur Res.	1970-1981	1.1 mil. tonnes @ 1.95% Cu, and 0.83 mil. Tonnes @8.47% Zn
Louvicourt	Louvicourt Twp.	Aur Res.-Noranda-Teck Cominco	1995-2005	6.0 mil. tonnes@ 1.7% Zn,2.90% Cu,26.3 gpt Au,0.02 opt Ag
Manitou	Bourlamaque Twp.	Louvem	1942-1979	3.6 mil. tonnes @ 1.0% Cu, 7.6 mil. Tons @ 4.52% Zn, .45% Pb, 1.13 gpt Au, 124.47 gpt Ag
EASTERN TOWNSHIPS				
Solbec	Stratford Twp.	Sullivan Mines	1962-1970	1.9 mil. tonnes @ 4.50% Zn, .69% Pb, 1.56% Cu, 0.58 gpt Au, 52.11 gpt Ag
Moulten Hill	Ascot Twp.		1944-1945, 1950-1953	317,000 tonnes @ 4.75% Zn, .86% Pb, .68% Cu, 63.42 gpt Ag
Hilton Mines	Shawville	Hilton	1872-1894, 1956-1976	
O'Deposit	Clinton Twp.	Clinton Copper	1974-1975	126,191 tonnes @ 2.5% Zn, 2.60% Cu, 0.41 gpt Au, 31.44 gpt Ag
Cupra-D'Estrie	Stratford Twp.	Sullivan	1967-1977	2.4 mil. tonnes @3.28% Zn, .54% Pb, 2.74% Cu
Huntingdon		Finneth	1954-1958	1.2 mil. tonnes @ 0.9% Cu, .03 opt Ag, .003 opt Au
Suffield	Ascot		1951-1955	589,000 tonnes @ 4.75% Zn, .86% Pb, .68% Cu, 63.42 gpt Ag
Weedon	Weeden Twp.	Sullivan Mines	1952-1959, 1969-1973	1.6 mil. tonnes @ .86% Zn, 2.33% Cu, 0.56 gpt Au, 4.37 gpt Ag
Harvey Hill	Leeds Twp.	Flying Cross	1974-1976	159,000 tonnes @ 1.25% Cu
TIMMINS CAMP				
Kam-Kotia	Robb Twp.	Cominco	1943-1944, 1962-1972	6.6 mil. tonnes @1.17% Zn, 1.15 Cu, 0.10 opt Ag
Kidd Creek	Kidd Twp.	Texas Gulf-Falconbridge-Xstrada	1966-Present	215.0 mil. tonnes @ 6.51% Zn, 0.26% Pb, 2.4% Cu, 8.20 opt Ag
Canadian-Jamieson	Godfrey Twp.	Cdn. Jamieson	1966-1971	0.816 mil. tons @ 4.2% Zn, 2.44% Cu
Jameland	Jamieson Twp.	Cdn. Jamieson	1966-1972	0.509 mil. tons @ 0.9% Zn, 0.99% Cu
MANITOUWADGE CAMP				
Geco	Gemmell Twp.	Noranda	1957-1994	52.8 mil. tonnes @3.7% Zn, 0.12% Pb, 1.8% Cu, 1.64 opt Ag to 1993
Wilroy	Gemmell Twp.	Lac Minerals	1957-1975	4.6 mil. tons @ 5.7% Zn, .1.25% Cu, 1.70 opt Ag
Will-Echo	Mapledoran Twp.	Lac Minerals	1957- 1977	3.8 mil. tons @ 3.87% Zn, 0.57% Cu, 1.96 opt Ag
Nana Creek	Gemmell/Mapledoran Twps.		-1976	251,236 tons @ 3.8% Zn, 0.83% Cu, 1.04 opt Ag
Winston Lake	Pays Plat Lake Area	Falconbridge Copper	1988-1989	3.1 mil. tonnes @ 14.5% Zn, 1.0% Cu, 0.93 opt Ag, .04 opt Au
Zenith	Pays Plat Lake Area	Zenmac	1966-1970	181,830 tons @ 16.5% Zn, 0.75 opt Ag, .02 opt Au
STURGEON LAKE CAMP				
Mattabi	Six Mile Lake Area	Mattagami-Noranda	1972-1991	19.0 mil. tons @ 8.5 % Zn, 1.05% Cu, 0.9% Pb, 3.31 opt Ag
Lyon Lk./Creek	Six Mile Lake Area	Mattagami-Noranda	1980-1991	3.2 mil. tons @ 8.7% Zn, 0.99% Pb, 1.26% Cu, 4.50 opt Ag
F Zone	Six Mile Lake Area	Mattagami-Noranda	1980-1984	0.4 mil. tons @ 9.5% Zn, 0.58% Pb, 0.64% Cu
Sturgeon	Six Mile Lake Area	Falconbridge-Copper	1974-	2.3 mil. tons @ 9.2% Zn, 1.21% Pb, 2.55% Cu, 5.17 opt Ag
South Bay	Dent Twp.	Selco Mining	1971-1980	1.64 mil. tons @ 9.7% Zn, 1.38% Cu, 1.75 opt Ag
Berens River	Setting Net Lake	Zahaffy Getty	1938-1948	560,000 tonnes @ 1.45% Zn, 4.94% Cu, 0.26 opt Au, 10.30 opt Ag
Central Hill	Matheson			600,000 tonnes @ 1.67% Zn, 2.05% Cu
Kingdon	Renfrew			900,000 tonnes @ 3.32% Pb

MANITOBA

MINE NAME	LOCATION	PRODUCER	MINE LIFE SPAN	PRODUCTION
FLIN FLON CAMP				
Flin Flon Mine	Flin Flon	Hudson Bay	1930-1992	19.2 mil. tons @ 4.13% Zn, 2.18% Cu, 2.5 g/t Au, 41.5 g/t Ag
Cuprus	Flin Flon	Hudson Bay	1948-1954	462,000 tons
Ghost	Flin Flon	Hudson Bay	1972-	
Western	Flin Flon	Hudson Bay	1978-	
Centennial	Flin Flon	Hudson Bay	1978-1983, 1985-1988	2.4 mil. tons @ 2.73% Zn, 1.45% Cu
Trout Lake	Flin Flon	Hudson Bay-Hudbay	1981-Present	9.9 mil. tons @ 5.41% Zn, 1.98% Cu
Callinan	Flin Flon	Hudson Bay-Hudbay	1990-Present	

MINE NAME	LOCATION	PRODUCER	MINE LIFE SPAN	PRODUCTION
Triple 7	Flin Flon	Hudson Bay-Hudbay	1999-Present	
West Arm	Schist Lake	Hudson Bay	1978-1985	1.6 mil. tons @ 1.42% Zn, 1.45% Cu
White Lake	Flin Flon	Hudson Bay	1971-1982	844,598 tons @ 4.63% Zn, 1.87% Cu
Spruce Point	Reed Lake	Hudson Bay	1986-1992	726,400 tons @ 4.4% Zn, 2.7% Cu
Don Jon	Flin Flon	Hudson Bay	1955-1958	79,000 tons @ 3.03% Cu
Schist Lake	Flin Flon	Hudson Bay	1954-1975	1.9 mil. tons @ 7.0% Zn, 4.21% Cu
North Star	Flin Flon	Hudson bay	1955-1958	242,000 tons @ 6.11% Cu
SNOW LAKE CAMP				
Chisel Lake	Snow Lake	Hudson Bay	1960-88, 1990-1994	7.3 mil. tons @ 10.5% Zn, 0.15% Cu
Chisel North	Snow Lake	Hudson Bay	-1980	2.7 mil. tons @ 9.0% Zn, 0.17% Cu
Anderson Lk.	Snow Lake	Hudson Bay	1970-1983	3.2 mil. tons @1.0% Zn, 3.47% Cu
Rod #1	Snow Lake	Hudson Bay	1962-1964	23,000 tons @ 4.5% Zn, 5.0% Cu
Rod #2	Snow Lake	Hudson Bay	-1984	645,000 tons @ 2.85% Zn, 6.15% Cu
Stall Lake	Snow Lake	Hudson Bay	1964-1984	6.3 mil. tons @ 4.8% Zn, 4.33% Cu
Photo Lake	Snow Lake		1995-	
Osborne	Snow Lake	Hudson Bay	1968-1983	3.4 mil. tons @ 1.52% Zn, 3.14% Cu
Spruce Point Lake	Snow Lake	Hudson Bay	-1993	2.0 mil. tons @ 3.18% Zn, 2.35% Cu
Ghost Lake	Snow Lake	Hudson Bay	1972-1988	606,000 tons @ 8.55% Zn, 1.27% Cu, 0.39 gpt Au, 0.45 gpt Ag
Dickstone	Snow Lake	Falconbridge	1970-1975	1.1 mil. tons @ 3.17% Zn, 2.42% Cu, .02 opt Au, .37 opt Ag
LYNN LAKE CAMP				
Fox	Lynn Lake	Sherritt-Gordon	1970-1985	14.6 mi. tons @ 1.91% Zn, 1.27% Cu
Ruttan Lk.	Leaf Rapids	Sherritt Gordon-Hudson Bay	1973-2002	40.0 mil. tons @ 1.36% Zn, 1.30% Cu
East, West	Sherridon	Sherritt-Gordon-Hudson Bay	1933-1950	7.7 mi tons @ 2.4% Cu, 2.0% Zn, 42.0 g/t Ag, 0.41 g/t Au

SASKATCHEWAN

MINE NAME	LOCATION	PRODUCER	MINE LIFE SPAN	PRODUCTION
Western Nucleur	Hanson Lake	Share Mines & Oils	1966-1969	147,300 tons @9.99 Zn, 5.83 Pb, .51 Cu, 0.31 opt Au, 4.47 opt Ag
Konuto Lake	Denara Beach	Hudson Bay-Hudbay	1997-Present	572,417 tons @ 1.16% Zn, 4.48% Cu, 1.99 g/t Au, 9.25 g/t Ag to 2000
FLIN FLON CAMP				
Flin Flon Mines	Flin Flon	Hudson Bay	1930-1992	43.2 mil. tons @ 4.17% Zn, 2.21% Cu, 2.5 g/t Au, 42.2 g/t Ag
Callinan North	Flin Flon	Hudson Bay-Hudbay	1997-Present	144,519 tons@4.17% Zn, 1.47% Cu, 1.64 g/t Au, 18.00 g/t Ag to 2002
Birch Lake	Flin Flon	Hudson Bay	1957-1960	278,690 tons @ 6.17% Cu, 0.1 g/t Au, 4.1 g/t Ag
Flexar	Flin Flon	Hudson Bay	1969-1972	305,940 tons @ 0.47% Zn, 3.75% Cu, 1.4 g/t Au, 8.9 g/t Ag
Coronation	Flin Flon	Hudson Bay	1960-1965	1.3 mil. tons @ 0.3% Zn, 4.25% Cu, 2.0 g/t Au, 4.8 g/t Ag

BRITISH COLUMBIA

MINE NAME	LOCATION	PRODUCER	MINE LIFE SPAN	PRODUCTION
Tulsequah Chief	Atlin	Cominco	1951-1957	
Samatosum	Adams Lake	Kerr-Inmet	1989-1992	554,837 tons @ .72% Cu, 774 gpt Ag
Goldstream	Revelstoke	Noranda-Goldstream	1982-1984,1991-1996	2.2 mil. tons @ 3.9% Cu, 11.92 gpt Ag
CAMPBELL RIVER CAMP				
HW, Lynx, Myra	Myra Falls	Western Mines-Breakwater	1966-Present	21.9 mil. tons @ 1.97% Cu, 1.46 gpt Au, 42.36 gpt Ag

BIBLIOGRAPHY

Prospecting is a thousand to one shot.

Karl Springer, 1932, in a letter to the federal Minister of Mines in response to heavy taxation.

Armstrong, C. *Moose Pastures and Mergers: The Ontario Securities Commission and the Regulation of Share Markets in Canada*, 1940-1980, Toronto: U of T Press, 2001.

Barnes, M. *The Scholarly Prospector: Don McKinnon*. Renfrew: General Store Publishing House, 2007.

________. *Looking Back COBALT*. St. Catharines: Vanwell, 2004

________. *Looking Back KIRKLAND LAKE*. St. Catharines: Vanwell, 2003.

________. *The Tri-Towns*. Cobalt: Highway, 2000.

________. *Great Northern Ontario Mines*. Renfrew: General Store Publishing House,2001

________. *Gold in Ontario*. Erin: Boston Mills, 1995.

________. *Great Northern Characters*. Renfrew: General Store Publishing House, 1995.

________. *Kirkland Lake on the Mile of Gold*. Erin: Boston Mills, 1994.

________. *Timmins: The Porcupine Country*. Erin: Boston Mills, 1995.

________. *Fortunes in the Ground*. Erin: Boston Mills, 1993.

________. *The Town that Stands on Gold*. Cobalt: Highway, 1978.

________. *Gold in the Porcupine*. Cobalt: Highway, 1976.

________. *Gold Camp Pioneer*. Cobalt: Highway, 1973.

Barr, E., and B. Dyck. B. *Ignace: A Saga of the Shield*. Winnipeg: Prairie Pub., 1979.

Carington, J. ed. *Risk Taking in Canadian Mining*. Toronto: Pitt Pub., n.d

Cassidy, George. *Arrow North: The Story of Temiskaming*. Cobalt: Highway Books, 1976.

Cork, E. *What Happened in Murdochville*. Lake Orton, Mich: private, 1999.

Cranstone, D. *A History of Mining and Mineral Exploration in Canada*. Ottawa: Natural Resources Canada, 2002.

________. *Outlook for Canadian Mine Production*, Mine Life and Ore Reserves. Ottawa: Natural Resources Canada, 2002.

Cruise,D., and A. Griffiths. *Fleecing the Lambs*. Vancouver: Douglas & McIntyre, 1987.

Cumming, J. *Diamond Drill Handbook*. Toronto: Smit & Sons, 1975.

De la Vergne, J. *Hard Rock Miner's Handbook*. Tempe, Arizona: McIntosh Engineering, 2000.

Deverell, J. *Falconbridge: Portrait of a Canadian Mining Multinational*. Toronto: Lorimer, 1975.

Drohan, M. *Scandals and their Aftermath*. Calgary: Chamir Foundation for Ethics, 2005.

Eckstrand,O., and W.D. Sinclair, et al. *Geology of Canadian Mineral Deposit Types*. Ottawa: Geological Survey of Canada, 1995.

Excell, J., and D. Haysom. *Mining Information for Aboriginal Communities*. Ottawa: Queen's Printer, 2006.

Finlay, V. *Buried Treasure: Travels through the Jewel Box*. London: Sceptre, 2006

Fraser, H. *A Journey North: The Great Thompson Nickel Discovery*. Thompson,Man.: Inco, 1985.

Frolick,V. *Fire Into Ice: Charles Fipke and the Great Diamond Hunt*. Vancouver: Rain Coast, 1999.

Geren,R., and B. McCullogh. *Cain's Legacy: The Building of the Iron Ore Company of Canada* Sept Isles: Iron Ore Co, 1986

Girdwood,C. et al. *The Big Dome*. Toronto: Cybergraphics, 1983

Goldie, R. *Inco Comes to Labrador*. St. John's NL: Flanker, 2005.

Goold, D., and A. Willis. *The Bre-X Fraud*. Toronto: McClelland & Stewart, 1997

Hanukla,M., ed. *The Discoverers: A 50 Year History of the PDAC*. Toronto: Pitt, 1982.

Hart, M. *Golden Giant*. Vancouver: Douglas & McIntyre, 1985.

______ *Diamonds*. Toronto: Penguin, 2001.

Hoffman, *A. Free Gold: The Story of Canadian Mining*. New York: Rinehart, 1947.

Howard, T. *Diamond Year*. Sandton, South Africa: Natal Witness Pub., 1996.

Hutchinson, B. *Fools Gold*. Toronto: Knopf, 1998.

Kupsch,W., Hanson,S. eds. *Gold and Other Stories*. Regina: Sask. Mining Assoc., 1986.

LeBourdais, D.M. *Falconbridge: The Sudbury Basin* 1928-1953. Toronto: Ryerson, 1953.

Lefolii, K. *Claims*. Toronto: Key Porter, 1987.

Longo, R. *Historical Highlights of Canadian Mining*. Toronto: Pitt Pub., 1973.

Longyear, E., and W. Eastman. *Longyear: The Mesabi and Beyond*. Gilbert, Minn.: 1984.

Lonn, G. *The Mine Finders*. Toronto: Pitt pub., 1966.

Martin, G. *Once Upon A Mine: A Story of Pre-Confederation Mines on the Island of Newfoundland*. Montreal: CIM spec. vol. 26, 1983.

_______. *Gesner's Dream: The Trials and Triumphs of Early Mining in New Brunswick*. Fredericton: CIMM-New Brunswick, 2003.

McDowall, D. *Steel at the Sault*. Toronto; U of T Press,1984.

McQueen, R.W. ed. *Proud Heritage: People and Progress in Early Canadian Geoscience*. St. John's NL: Geological Association of Canada, 2004.

McNish, J. *The Big Score: Robert Friedland, Inco, and the Voisey's Bay Hustle*. Toronto: Doubleday, 1998

n.a. *Drilling: The Manual of Methods, Application and Management*. Australia: Australian Drilling Industry Training Committee Manual, 1996.

Pain, S. *Three Miles of Gold*. Toronto: Ryerson, 1960.

_______. The Way North. Toronto: Ryerson, 1964.

Parres, J., and M. Jackson. *The Nor-Acme Gold Mine Story*. Thunder Bay: private, 2006.

Patrick, K. *Perpetual Jeopardy: The Texas Gulf Affair*. Toronto: Collier Macmillan, 1972.

PDAC. "Exploration and Development Highlights." Toronto: an annual. n.d.

PDAC. "Presentation to Federal Government in Ottawa." Toronto: PowerPoint show, April 24, 2006.

Ross, A. *The Traders*. Toronto: Collins, 1984.

Schulman, M. *The Billion Dollar Windfall*. Toronto: McGraw Hill, 1969.

Seavoy, R. *A New Exploration of the Canadian Arctic*. Surrey, B.C.: Hancock House, 2002.

Shaw, M. *Canadian Portraits: Geologists and Prospectors*. Toronto: Clarke,Irwin, 1958.

Smith, P. *Harvest from the Rock: A History of Mining in Ontario*. Toronto: Macmillan, 1986.

Sudol, S. *Claiming Our Stake*. Sudbury: City of Sudbury, Sept. 2006.

Swift, J. *The Big Nickel: Inco at Home and Abroad*. Kitchener: Between the Lines, 1977.

Taylor, B. *Steep Rock*. Atikokan: Quetico Pub., 1978.

Taylor, G. *The History of Mining in British Columbia*. Saanichton, B.C.: Hancock House, 1978.

Telford, W. et al. *Applied Geophysics*. Cambridge, UK: Cambridge UP, 1976.

Wells, J. *The Pez: The Manic Life of the Ultimate Promoter*. Toronto: Macfarlane, Walter, Ross, 1991.

Werniuk,G. ed. "PDAC 75". Toronto: Northern Miner, 2007.

Wilson, L. *Chibougamau Venture: The Story of a Quebec Mining Field*. Montreal: Chibougamau Publishing, 1952.

Zaslow, M. *Reading the Rocks: The Geological Survey of Canada*. Ottawa: Macmillan, 1975

Magazines, etc

"Mr. Exploration." The Northern Miner Magazine, March 1989.

There are many prospectors and others who have contributed much to mining and the Canadian economy and have no mention of their work anywhere in print. Ecclesiastics Chapter 44, verse 9 echoes their epitaph:

"There are those of them that have left a name behind them, that their praises might be reported. And some there be which have no memorial."

ACKNOWLEDGEMENTS

In the event of an emergency over water please hang on to your cushion as it will help you to swim to shore. In such an event, keep the cushion with our compliments.

Sign noticed on a bush plane.

This book could not have been attempted without the vision and technical expertise of John Harvey, former V.P. Exploration for Noranda, the goodwill and material backing of Robert Cudney of Northfield Capital, perhaps better known as a principal in Gold Eagle Mines at Red Lake, and the willingness of Dr. Jim Franklin, late chief scientist of Natural Resources Canada, to contribute from his experience and wealth of knowledge of geology and the economics of the mining industry.

Staff at the Prospectors and Developers Association of Canada who smoothed the way were Tony Andrews, Saley Lawton, MaryAnn Mihychuk, and Laura Clinton of Mining Matters.

David Barr, an "original" at Voisey's Bay and Galore Creek, gave B.C. hospitality and numerous insights.

Jim Kelly kindly gave permission to use his cartoons on the mining industry and created another for this book. Dennis Newman donated the use of his sketches of head frames. Many of these magnificent structures across Canada have been demolished and such images help us to recall their striking outlines against the northern skies.

Mining Exploration people of Queen's University who hosted the author at a mining seminar are David Love, Charles Pelley, Michael Doggett, Gema Ribeiro Olivo and Boyd Davis.

Helpful folks at the Royal Canadian Mint were David Madge, Rod Beauprie, and Debbie Borden.

Those in government and also QUANGOs (Quasi Autonomous Non Government Organizations) who helped the author were Robert Clarke, Arlene Drake and Mark Pilkington (NRCan), Pam FitzMaurice (N.W.T.), Bakes Mitchell (Mines Economics Group), Pam Sangster and Sean MacLean (ON MNDM), Dr. Steve McCutcheon and Don Carroll (New Brunswick), Kelly Mahoney (N.W.T.), Dave MacDougall (Sask), Karen Dare (Geol. Assoc. Canada), Richard Bennett (Sask), Richard Wardle, David Leonard, Sean O'Brien and Norm Mercer (Newfoundland), Leon Carbonneau (Quebec), Kelly Mahoney (N.W.T.), Scott Cairns (N.W.T.), Ryan Silke (N.W.T.), Colin Hunt (Can. Nuclear Assoc), John Scott and Ann Wilson (OMNR), Tom Schroeter (BC EMPR), Karen Costello (DIAND Nunavut), Dave Guindon (OMNR), David MacDougall (Sask. Industry), Pierre Verpaelst (Quebec Mines Ministry), John Scott and Gerry White (ONDM), Tracy Burton (ONDM), Oliver Bonham (Can. Council Professional Geoscientists), Diane Edmond (Yukon Geological Survey),David MacDougall (Sask. Industry and Resources).

Helpful library assistants were Vicki Fraser (Haliburton), Gwen Schmidt (Saskatoon), Andrew Martin (Vancouver), Jennifer Knowles (Yellowknife),

Generous people in private business were Brent Peters, Clayton Parsons and Chantelle Melanson (Major Drilling), Al Workman (Watts,etc), Peter Dunsford (Teck Comminco), Vince Borg and Bill Gray (Barrick), Roberta Voyer (Peregrine), James Wong (San Gold), Brian Hester, Carl Huston, Rhylin Bailie, Christine Daniels, Donna Yashimatsu, Karen Yun, Ray Mlait, Nick Thomas (Stornoway), Pat Roberts (Northern Miner), Aline Cote (Xstrata), Christine Ogryzlo (Smithers Exploration), Scott Herr (Inmet Mining), Brent Jones (Century Mining), Glen Davidge (Xstrata Kidd), Bill Gray (Barrick), Genevieve Desmarais (Fugro), Tom Hoefer and Doug Ashbury (Diavik), Adriana Peretz (Nova Gold),Abraham Drost (Sabina Silver), Terry Piorun (Adanac), Virginia Barraga (Universal Wing), John Davies (Mintec), Bruce Mitton (Cornerstone Res.), Brooke Yeates, Cory McPhee (CVRD Inco), Chad Calaiezzi (Expedition Helio),Tim Leriche and Trina Thompson (ICO), Annie Stavridis (De Beers), David Con stable (FNX),Marise Milan (Boart), Alison Ascott (Rockware) and David Fountain (Fugro).

Many thanks to Fran Harvey who typed the Excel files and to John Stevens who edited the manuscript with his usual kindness and panache.

ABOUT THE AUTHOR

Michael Barnes has been a teacher and school principal, a correspondent for the *Globe and Mail*, a CBC freelancer, newspaper columnist for Southam and Thomson and an author of books on Canadian resources, particularly mining, northern communities and police work, mostly involving the Ontario Provincial Police (OPP).

Some of his books on the mining industry include *The Scholarly Prospector: Don McKinnon, Looking Back COBALT, Looking Back KIRKLAND LAKE, Great Northern Ontario Mines, Gold in Ontario, Great Northern Characters, Kirkland Lake on the Mile of Gold, Timmins: The Porcupine Country* and *Fortunes in the Ground*. An early book for children was *Monster from the Slimes*. His next book will be *RED LAKE: Golden Treasure Chest*.

Michael Barnes is a member of the Order of Canada and an Honorary Inspector in the OPP. He lives in Haliburton, Ontario.

More details of his work may be found in the Members pages of The Writers Union of Canada website.

Michael Barnes may be contacted at *michaelbarnes53@hotmail.com*.